A Naturalist's Guide to the Texas Hill Country

Number Fifty

W. L. Moody Jr. Natural History Series

A|M *nature guides*

A Naturalist's Guide to the Texas Hill Country

Mark Gustafson

Texas A&M University Press
College Station, Texas

♾ This paper meets the requirements of ANSI/NISO Z39.48–1992
(Permanence of Paper).
Binding materials have been chosen for durability.
Manufactured in China through Martin Book Management

LIBRARY OF CONGRESS CATALOGING-IN-PUBLICATION DATA

Gustafson, Mark, 1967– author.
  A naturalist's guide to the Texas Hill Country / Mark Gustafson. — First edition.
    pages cm — (W.L. Moody Jr. natural history series ; number fifty)
  Includes bibliographical references and index.
  ISBN 978-1-62349-235-9 (flexbound (with flaps) : alk. paper)—
  ISBN 978-1-62349-236-6 (ebook)
  ISBN 978-1-64843-331-3 (paper (with flaps) : alk. paper)
1. Natural history—Texas—Texas Hill Country. I. Title. II. Series: W.L. Moody Jr.
natural history series ; no. 50.
  QH105.T4S87 2015
  508.764—dc23

2014031477

Contents

Acknowledgments

I WOULD LIKE TO THANK my family and friends for their support of this project. I especially want to thank my son, Benjamin, for accompanying me on so many of the field excursions that were necessary for the book. I thank Texas Lutheran University for providing a sabbatical leave that enabled me to complete a substantial portion of the book. Thanks also to the Weston Ranch Foundation for summer research support that contributed to my understanding of the Hill Country. My colleagues Alan Lievens, Lorne Davis, and Judy Hoffman read chapters and provided many useful comments. Robert Egan and Robert Lücking aided in the identification of the fungus *Robergea albicedrae*. I am very grateful to the following photographers for providing photographs for the book: Hannah and Craig Meddaugh, Dustin and Lindsey Wyatt, Chris Harrison, Forrest M. Mims III, Tripp Davenport, Carolyn Whiteside, Sam Stukel, Gary Nafis, Susan Sander, US Fish and Wildlife Service, Texas Parks and Wildlife Department, and Lower Colorado River Authority. Thanks to Shannon Davies, editor-in-chief, and Patricia Clabaugh, project editor, at Texas A&M University Press, for all their assistance during the publication process. This book is dedicated to Wendi, Kiersten, and Benjamin and to my mother and father.

A Naturalist's Guide to the Texas Hill Country

Introduction

THE HILL COUNTRY IS PERHAPS the best-loved region in Texas. It is known for its beautiful displays of spring wildflowers, scenic vistas, clear and cool streams, and quaint small towns. It is also a region with a surprising diversity of life, including a large number of species found nowhere else in the world. It is a crossroads of life, with species present that are typical of the desert to the west, the grasslands to the north, the forests to the east, and the tropics to the south. Although much of it is lightly used by humans at present, it has profoundly changed since the early settlers arrived and is now undergoing major changes again as large metropolitan areas encroach from its eastern edge.

The term "Hill Country" is usually used to describe the eastern part of the Edwards Plateau. This huge limestone plateau covers much of the central part of Texas, and the eastern and southern portions of it have been severely eroded to form the Balcones Canyonlands. Some people like to say that the Hill Country is really Valley Country, because the hills were not pushed up like mountains but instead were left in place as erosion carved out valleys between them. In the northeastern part of the plateau is a region of ancient granite rocks called the Llano Uplift, which is also usually considered part of the Hill Country. The Edwards Plateau Ecoregion is another name for the biologically distinctive area that includes the Hill Country and the western part of the Edwards Plateau.

The western part of the Edwards Plateau is less steeply eroded and therefore is usually not included as part of the Hill Country. This area is also more arid and desertlike and is less visited. This book focuses on the 19-county region shown on the map; however, most of the species found in the Hill Country are also found in nearby areas.

A Naturalist's Guide to the Texas Hill Country provides an introduction to the geology, rivers and lakes, plants, and animals of the Hill Country. As users of this book explore the Hill Country, they will gain a better understanding of how all of these natural features are related.

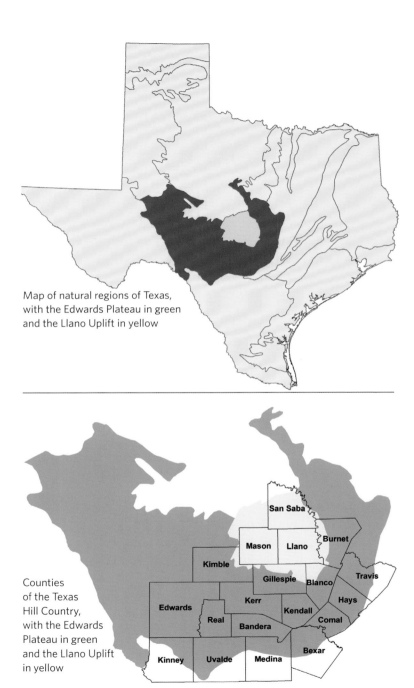

Map of natural regions of Texas, with the Edwards Plateau in green and the Llano Uplift in yellow

Counties of the Texas Hill Country, with the Edwards Plateau in green and the Llano Uplift in yellow

San Saba

Mason   Llano   Burnet

Kimble

Gillespie   Blanco   Travis

Edwards   Kerr   Kendall   Hays

Real   Bandera   Comal

Kinney   Uvalde   Medina   Bexar

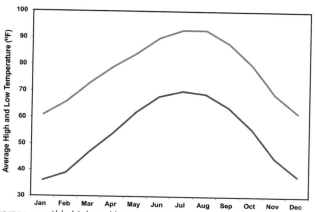

Average monthly high and low temperatures for Fredericksburg, located in the middle of the Hill Country

Because hundreds of species of plants and animals are present in the region, only the most common and unique species can be included here. However, these are the species most likely to be encountered.

The Hill Country is a year-round destination. Fall, winter, and spring rains bring the spectacular spring wildflower displays that every photographer wants to capture (March to May). The weather is pleasant in the spring, with cool temperatures during the day. Summers are hot and usually dry (June to September). In summer, it is best to take a hike early in the morning and then spend the afternoon swimming, tubing, or kayaking in one of the spring-fed rivers. Fall usually does not arrive until October, when cool nights make for exceptional camping and hiking. Fall colors in late October and early November bring hundreds of people every day to Lost Maples State Natural Area. Winter is quiet, and most of the vegetation is brown, but the daytime temperatures are usually pleasant for outdoor activities. Occasional cold fronts, called "blue northers" by Texans, can rapidly bring freezing temperatures and even snow. Caves in the Hill Country have temperatures ranging between 65°F and 75°F year-round and thus are a good way to escape the heat of a hot summer afternoon or the cold of a winter day. Typical temperatures for the Hill Country are shown in the graph above.

Enchanted Rock

Geology of the Hill Country

The Hill Country exists as a unique region in Texas because of its geologic history. The formation of the limestone rocks that make up its modern-day hills has resulted in its topography, abundance of caves, and a giant aquifer that provides clear, cool water to its springs and streams. Even many of the plants that grow in the Hill Country are there because of the limestone.

GEOLOGIC HISTORY

Geologic history extends back to the formation of the earth approximately 4.5 billion years ago. The oldest rocks exposed in the Hill Country are igneous rocks (solidified from molten rock, so they do not contain fossils) of the Llano Uplift, which are about 1 to 1.3 billion years old. The granite that makes up Enchanted Rock is part of these Precambrian rocks. Granite from this ancient formation has been quarried and was used in constructing the Texas State Capitol in Austin. Enchanted Rock itself is only a small part of a huge granite formation called the Enchanted Rock Batholith. Inks Lake State Park is another good place to observe this ancient granite.

During the Paleozoic era, sedimentary limestone rocks formed that are now exposed in a few parts of the Hill Country. During this era, shallow oceans covered the region. As sediments and dissolved

Above: Limestone layers in the Hill Country at Canyon Lake Gorge

Left: Marine bivalve (clam) found in limestone

minerals were carried from land to the sea, they settled out or precipitated in the warm, shallow water. These rocks are exposed in the riverbed at Pedernales Falls State Park, where they have been tilted and cracked, creating a series of small waterfalls. These rocks are part of the Marble Falls Limestone and were formed during the Pennsylvanian period about 300 million years ago. Longhorn Cavern was formed in the limestone from the Ordovician period, about 500 million years ago.

Most of the Hill Country consists of limestone rock formed much later, during the Cretaceous period of the Mesozoic era. During parts of this period, from 138 to 65 million years ago, most of Central Texas was under a shallow ocean. As in the Paleozoic, sediments and dissolved minerals were deposited on the ocean floor. Organisms living in the sea died and sank to the bottom, where they were buried in the sediments and many became fossilized. Fossils of marine animals such as clams, snails, and sea urchins are the most common large fossils found in the limestone. The main rock layers from this time period are called the Glen Rose Limestone and the Edwards Limestone. The Glen Rose Limestone is older and therefore lies underneath the Edwards Limestone where both are present. For example, in Lost Maples State Park, the hilltops are Edwards Limestone but the canyon walls and valley bottoms are Glen Rose Limestone. The Edwards Limestone is extracted for use as road base from several quarries along Interstate Highway 35, from San Antonio to Georgetown. Note that these rocks, which were formed below sea level, were later uplifted by about 2,000 feet to form the Edwards Plateau (uplift was complete by 10 to 20 million years ago). At this time, the Balcones Fault Zone was also created on the southeastern edge of the Hill Country, where the rock layers are broken on the edge of the plateau. Ever since, the rock of these limestone layers has been eroding away, gradually washing down toward the Gulf of Mexico.

Because most of the rocks of the Hill Country were formed during the Mesozoic era, or age of dinosaurs, those that were in shallow waters or mudflats near the shore were potentially walked across by dinosaurs. Although dinosaur bones are not common in the Hill Country, dinosaur tracks have been found throughout the region, most of them on private land. The best place to view dinosaur tracks in the Hill Country is the Heritage Museum of the Texas Hill Country, located on FM 2673 on the south side of Canyon Lake. These tracks are in the Glen Rose Limestone, which is the same formation as the well-known tracks found north of the Hill Country at Dinosaur State Park near Paluxy. Two main types of tracks occur in Central Texas. The first is the sauropod track, usually a large rounded track, which was made by large, four-legged, long-tailed herbivores, most likely *Pleurocoelus.* The other type, a tridactyl track, has three distinct toes and was made either by herbivores or carnivores. Tracks of *Acrocanthosaurus,* a huge carnivorous dinosaur related to *Tyrannosaurus,* have been found in the Hill Country.

Dinosaur track

There are also fossils from the more recent ice ages, part of the Pleistocene epoch of the Cenozoic era. This period lasted from about 2.5 million to 12,000 years ago. The nearest ice sheets were in Illinois, and the climate in Texas was cool and wet during this period. Large mammals roamed the landscape of the Hill Country, and occasionally their bones were preserved in caves or sinkholes. Friesenhahn Cave near San Antonio (on private land) contains the remains of more than 30 species of vertebrate animals from the Pleistocene. These include saber-toothed cats, woolly mammoths, peccaries, turtles, and sloths. The Waco Mammoth Site, located to the northeast of the Hill Country, contains the bones of more than 20 mammoths that were probably killed by a flash flood and buried in mud about 68,000 years ago.

The soils of the Hill Country are the product of its geology and vegetation, as well as the impact of humans. The soils on the hills tend to be thin, due to the sparse vegetation combined with occasional heavy rains that lead to erosion. Limestone rock is often exposed or just below the soil surface. The lack of deep soil limits the growth of plants, although some plants can extend their roots into the water table below the topsoil. The soils in the river valleys are much deeper and richer, due to deposition of soil that has been eroded from the hillsides. Most of the agricultural production and most towns are thus located along the river valleys.

KARST TOPOGRAPHY AND CAVES

Because most of the Hill Country is composed of limestone that has been eroding and dissolving for millions of years, this area is a region of karst topography, an area of limestone that now contains many cavities. As raindrops pass through the atmosphere, they absorb carbon dioxide (CO_2) from the atmosphere. When carbon dioxide dissolves in water, it forms carbonic acid, which reacts with the calcium carbonate of limestone and dissolves holes in the rock. These holes can be seen in many of the limestone outcrops in the Hill Country. As water percolates down through cracks in the limestone, it dissolves rock and forms larger and larger cavities that become caves, which are widespread throughout the limestone of the Hill Country.

Deeper below the surface, water collects in the cavities, forming aquifers. Wherever the cavities connect to one another, water can flow through the aquifer. As water accumulates in the aquifer, it creates pressure on the water below it due to its weight. This pressure creates the flow of springs in places where the aquifer reaches the surface. Springs are abundant throughout the Hill Country, but the largest ones are located along the Balcones Fault Zone from San Antonio to Austin. Comal Springs, San Marcos Springs, and Barton Springs are the largest and best known of these outlets.

In contrast to the springs where water flows out of the aquifers, there are also recharge zones where water flows in. These are located in areas where cracks in the limestone allow water to flow down into the aquifer. When streams cross these areas, their flow will decrease as water drains down into the ground (in other words, rather than getting larger downstream as most streams do, the stream will actually be smaller downstream). In many cases, the stream actually disappears underground and may then reappear as springs farther downhill.

One of the best ways to experience the geology of the Hill Country is to take a tour of a cave. The following Hill Country caves are open to the public:

> Cascade Caverns, Boerne:
> home to an endemic species, the Cascade Caverns
> salamander; daily tours

Cave entrance in San Antonio

Cave without a Name, Boerne:
    National Natural Landmark; daily tours
Colorado Bend State Park, Bend:
    Gorman Cave; wild cave tours by reservation only
Inner Space Caverns, Georgetown: daily tours
Kickapoo Cavern State Park, Brackettville:
    wild cave tours by reservation only
Longhorn Cavern State Park, Marble Falls:
    National Natural Landmark; cavern is in Paleozoic
    limestone; daily tours
Natural Bridge Caverns, New Braunfels:
    National Natural Landmark; largest commercial caverns
    in Texas; daily tours
Wonder Cave, San Marcos:
    cave formed by an earthquake; daily tours

Water in the Hill Country

The Hill Country's location in the middle of Texas gives it a moderately arid climate. Annual rainfall ranges from 30 to 35 inches in the eastern Hill Country to 20 to 25 inches in the western part, which is farther from the Gulf of Mexico. Rainfall is usually highest in the spring and fall. Periods of extended drought are common, with little rain for months. The high temperatures of summer cause rapid evaporation of water from the soil surface. Trees may be able to obtain water from pools of water trapped in the rocky soil. Most of the native plant species are either well adapted to drought or live in the valley bottoms where water is more available.

SPRINGS AND STREAMS

The patterns of flowing water in the Hill Country are the result of the karst topography. Water easily flows into the ground through the many cracks and crevices in the limestone, but there are also many springs where these cracks allow the water out of the ground. Several major rivers begin in the Hill Country and then flow down across the coastal plain to the Gulf of Mexico. Due to the convergence of cool air from the north with moist air from the Gulf of Mexico, very heavy rainfall occasionally occurs in the Hill Country. This combines with the rapid runoff from sparsely vegetated lands and steep topography to create a region that is sometimes called Flash Flood Alley. Rainfalls up to 48 inches in a few days have created destructive floods for all of the cities along the southeastern edge of the Hill Country, as well as those on the coastal plain downstream. Even small amounts of rainfall can create local flash floods due to the thin soils and sparse vegetation that allow water to quickly run off into stream channels.

AQUIFERS

Beneath the arid landscape of the Hill Country lies another world, one filled with water. The porous nature of the limestone plateau has allowed huge aquifers to form underneath the Hill Country. These are somewhat like gigantic underground lakes, but they mostly fill small, interconnected spaces in the rock, not open caverns. The largest aquifers are the Trinity and Edwards Aquifers, which are combined in places. Water enters the aquifers through recharge zones, where gravity pulls the water down to the aquifer level. The aquifers reach the surface in the artesian zone, where water pressure pushes water out, forming some of the largest springs in the southwestern United States.

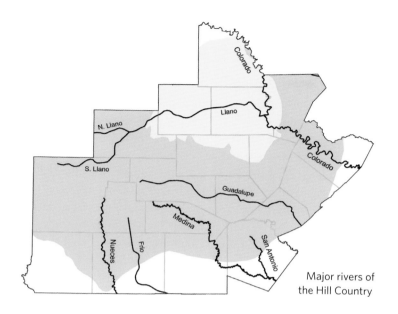

Major rivers of
the Hill Country

These include Comal Springs in New Braunfels, San Marcos Springs in San Marcos, and Barton Springs in Austin. In addition, thousands of wells have been drilled into this region to supply water for agriculture, cities, and industries. The city of San Antonio gets nearly all of its water from the Edwards Aquifer.

Despite the vast amount of water in the Edwards Aquifer, people can use only a small portion because of the water pressure. As water is pumped out of the aquifer, the level to which the water fills the spaces in the ground drops. As this happens, it exerts less pressure on the water below, reducing the force that causes the springs to flow. When it is not raining, the springs in New Braunfels and San Marcos are the primary sources of water for the Guadalupe River. Thus, depleting the aquifer below the level at which the springs flow would not only cause the extinction of the unique species that live there; it would also affect everyone downstream, including cities such as Seguin, Gonzales, Cuero, and Victoria. In addition, aquatic ecosystems downstream would be affected by reduced stream flow, including the estuaries where the Guadalupe meets the bays of the Gulf of Mexico. These estuaries are breeding grounds for many species of fish, as well as the winter home to the endangered Whooping Crane. A series of legal battles have been waged over the use of water from the Edwards Aquifer, with protection of the endangered species the main legal force that has prevented depletion of the aquifer to levels that would dry up the springs. The modern system of aquifer protection has not been completely tested, though. During the 1950s, a multiyear drought led to the drying of Comal Springs. If a similar drought occurred now, with millions more people dependent on the aquifer, the effects would be severe.

Human use of the water from the aquifer and the rivers of the Hill Country is regulated by complex state laws that do not necessarily recognize the interconnections of the water cycle. Surface water is generally owned by the state, and users must have legal permission to extract water from rivers. Groundwater is regulated under the "right of capture," which allows landowners to use as much water from below their land as they need. However, in recent years the right of capture has become limited in many areas of Texas through the formation of groundwater conservation districts, which can regulate pumping from groundwater in their region. Most of the Hill Country is now within various groundwater conservation districts or is under the regulation of the Edwards Aquifer Authority.

RESERVOIRS

Reservoirs have been created on many of the Hill Country rivers to prevent floods, store water, and provide recreational opportunities. A series of five dams form the Highland Lakes system on the Colorado River upstream from Austin. The table shows the large reservoirs, which were mostly constructed during the 1930s and 1960s. In addition, there are many small dams that create small impoundments along the rivers, but these play little role in flood control because the water easily flows over the dams after heavy rains. The upper parts of the Guadalupe, Llano, Frio, and Nueces Rivers are known for their scenic nature and are popular for kayaking, canoeing, and tubing.

AQUATIC LIFE

The streams and aquifers of the Hill Country have probably been here in some way for millions of years, and this has enabled a unique set

Major reservoirs of the Texas Hill Country

Reservoir	Surface area (acres)	Maximum depth (feet)	River
Buchanan	22,211	132	Colorado
Travis	18,622	190	Colorado
Canyon	8,308	125	Guadalupe
Lyndon B. Johnson	6,449	90	Colorado
Medina	5,426	152	Medina
Austin	1,599	75	Colorado
Inks	831	60	Colorado
Marble Falls	611	60	Colorado

Note: Surface area and depth are often much lower in periods of drought.

of species to evolve in these habitats. Endemic species are those that exist in only a small area and usually have evolved in that area. There are dozens of endemic aquatic species in the Hill Country, especially among the less mobile animals such as small fish, salamanders, and small insects. These factors make the Edwards Aquifer and its springs the most unique aquatic habitat in Texas. The following species of fish are all endemic to the Edwards Plateau:

Fountain darter	(*Etheostoma fonticola*)
Greenthroat darter	(*Etheostoma lepidum*)
Largespring gambusia	(*Gambusia geiseri*)
San Marcos gambusia	(*Gambusia georgei*)
Guadalupe bass	(*Micropterus treculi*)
Widemouth blindcat	(*Satan eurystomus*)
Toothless blindcat	(*Trogloglanis pattersoni*)

The springs, streams, and underwater cave habitats of the Hill Country are also home to a set of closely related brook salamanders (genus *Eurycea*). There are currently 13 species, but most of these were only recently identified through genetic studies, so further study is likely to increase the number of recognized species. Many are found only in a single system of springs or caves.

In addition, Texas wild rice (*Zizania texana*) is an endemic plant that lives only in the San Marcos River near the springs. Three invertebrate species are also unique to the springs in New Braunfels and San Marcos:

Comal Springs riffle beetle	(*Heterelmis comalensis*)
Peck's Cave amphipod	(*Stygobromus pecki*)
Comal Springs dryopid beetle	(*Stygoparnus comalensis*)

In summary, the waters of the Hill Country, based on underground aquifers, sustain an area of high biological diversity. Human use of the water and changes in the landscape above these aquifers have resulted in many of the species in this area being designated as endangered. Without abundant clean water flowing out of the ground, the Hill Country would not be so favored by both human and animal life.

Human History and Influence on the Land

In addition to the geologic, climatic, and biological influences on the Hill Country, humans have had profound impact on the land over the past 15,000 years. This influence is especially evident in the vegetation cover in the modern-day Hill Country.

PALEOINDIANS

Near the end of the last ice age, about 15,000 years ago, the first humans crossed from Asia to North America and started spreading throughout the continent and into South America. The earliest-known artifacts in the Edwards Plateau region were recently discovered in Williamson County on the northeast side of the Hill Country. This site, containing artifacts known as the Buttermilk Creek Complex, has clear evidence of stone tools that date from 13,200 to 15,500 years old. Thus, people have been living in the Hill Country since shortly after their arrival in North America.

The Clovis culture, which was widespread in North America about 13,000 years ago, also left evidence in the Hill Country. The distinctive Clovis points have been found at several sites in the Hill Country. These people were a hunter-gatherer culture, and there is evidence that they hunted mammoths and mastodons. This was perhaps their largest influence on the Hill Country of today, as the extinction of these large mammals was likely due at least in part to hunting by Paleoindians. Climate change as the ice age ended may also have played a role. However, even if these huge animals had survived the Paleoindians, it is doubtful they would have survived the successive waves of people who followed, up to the present day. Following the Clovis period, other groups of Indians lived in Central Texas, including the Tonkawa.

LIPAN APACHES AND COMANCHES

In the 1600s and 1700s, Apaches moved into the southern plains of Texas from farther north. They were originally from Canada. By the 1730s, Lipan Apaches were well established in the Hill Country along the Pedernales, Llano, and San Saba Rivers. However, by the end of the 1700s, they too were displaced as the Comanches moved into the region. The Comanches dominated the plains and hills of Central Texas until the mid-1800s. Fighting between the Comanches and white settlers who started invading their land in the 1800s was violent in northern Texas and slowed the spread of the settlers. By 1860, most of the Texas Indians, as well as the buffalo they depended on, were gone. Many of the Indians died as a result of the spread of diseases such as smallpox, to which they had no immunity. The last Comanches surrendered to the US Army in 1875 and moved to a reservation in Oklahoma.

The influence of Indians on the landscape of the Hill Country is difficult to determine, but it is likely that they contributed to suppression of shrub and tree growth when they set fires. Fires were used in hunting game as well as to create open prairie that was more productive for large grazing mammals such as buffalo.

EUROPEAN AMERICANS

Spanish explorers were the first Europeans to travel through the Hill Country. Spanish missions were established along the southern edge of the Hill Country in the early to middle 1700s, including those in San Antonio, which are preserved in the San Antonio Missions National Historical Park. Other missions were founded at San Marcos, New Braunfels, Austin, near Menard (on the northern side of the Hill Country), near Camp Wood, and near Montell. Most were in existence for only a few years.

During the 1830s and 1840s, the first settlers from the eastern United States and from Germany started moving onto the southeastern edge of the Hill Country, where the springs flow out of the aquifer. Austin was founded in 1839 as the capital of the Republic of Texas, which had been formed three years earlier. New Braunfels and Fredericksburg were founded in 1845 and 1846, just before Texas became a state, by German members of the Adelsverien, the Society for the Protection of German Immigrants in Texas. Settlement proceeded first along the main rivers, such as the Comal, Guadalupe, and Pedernales.

Early settlers cleared many of the large trees, especially along the rivers. Kerrville was founded by a settler from Kentucky who used the huge bald cypress trees along the Guadalupe River to make shingles. Floodplains were cleared for farming. Because of the rocky soil and steep slopes, much of the Hill Country was unsuitable for farming but was useful for grazing, especially goats and sheep.

MODERN ERA

The peak numbers of grazing animals occurred in the first half of the twentieth century. Stocking rates equivalent to about one cow (or five to six sheep or goats) per 5 to 7 acres were common on Hill Country ranches. Since then, the numbers of animals have declined dramatically in the Hill Country, with one cow per 20 to 30 acres a typical stocking rate. Much of this decline was likely due to reduced productivity of the land as a result of overgrazing. Overgrazing also led to less fuel for fires, which combined with fire suppression led to the encroachment of junipers ("mountain cedar") onto grazing land, leading to further declines in ranch productivity. However, white-tailed deer were favored by this transition to woody vegetation in the mid-twentieth century, and hunting leases soon became a way to earn income from the land. In addition, a wide variety of exotic game species such as axis deer were introduced, and trophy hunting became another source of income.

As agriculture and settlement spread through the Hill Country, several species of mammals were hunted to extinction in the late 1800s and early 1900s. The largest was the American bison, which previously grazed in large numbers and may have played a role in maintaining the prairie openings among the woodlands. The other species that were extirpated were predators: gray wolves, red wolves, ocelots, black-footed ferrets, and black bears (which are omnivores). At the time, all predators were considered a threat to livestock and humans, and their role in ecosystems was not understood. More recently, biologists have shown that predators are important in regulating the abundance and behavior of herbivores and thus the abundance of vegetation in ecosystems. It is likely that wolves especially played an important role in controlling deer numbers in the Hill Country.

There is much discussion in the Hill Country about the changes in vegetation that have occurred since European settlement began. The conventional view is that this area, which was once covered by prairies, has now been taken over by shrubland, especially Ashe juniper. Several studies of the historical record show that the Hill Country was not open grassland but a mosaic of grasslands, savannas, woodlands, and forests. Humans altered all of these habitats through grazing, suppressing fires, and cutting trees. Overall, it seems most likely that shrub cover has increased in some areas and decreased in others. For example, on the Kerr Wildlife Management Area, grassland area has decreased relative to that of historical surveys. At Camp Bullis north of San Antonio, it appears that the modern mix of half grassland and savanna and half woodland and forest is fairly similar to landscape of the mid-1800s. In the river valleys, forests have decreased due to clearing of fields for farming.

One of the most controversial topics in the Hill Country is the effects of juniper encroachment on the aquifers and flow of springs. A conventional wisdom has developed that Ashe juniper sucks water out of the ground, and that if landowners clear juniper, then aquifers will fill up and springs will start flowing out of the rocks. There is some evidence from local studies that spring flow does increase after juniper clearing and proper livestock management to prevent overgrazing. The Selah, Bamberger Ranch Preserve near Johnson City is a well-known case in which a severely overgrazed ranch was restored to healthier vegetation conditions, and spring flow increased. Some people have extended results such as this to suggest that ridding the Hill Country of junipers will increase the flow of springs throughout.

Scientific studies have shown mixed results when juniper was cleared, with some showing increased spring flow and others showing no change in spring flow. It seems most likely that large-scale clearing of all the cedar would likely not result in substantially increased aquifer levels or spring flow. In fact, historical data suggest that spring flow has actually been increasing since the 1950s during the period of juniper encroachment, when the ranches of the Hill Country started to end overgrazing. Overgrazing led to loss of vegetation, which increased runoff and thus reduced infiltration of the water into the ground. As the vegetation recovered, including the junipers, the ground began absorbing more of the precipitation. Careful studies of the root systems of junipers and oaks in the Hill Country suggest that they do not tap deep water from the aquifer but instead use shallow pools of water in the rocky soil. The soils and groundwater of the Hill Country are complex, and it seems that the effects of juniper clearing will depend on the local conditions, such as rainfall patterns, and may change through time as the vegetation changes after clearing. Because of the many endangered species that are dependent on the flow of water from the aquifer, research on how this complex system works will surely continue.

Urban and Suburban Development

The most significant current threat to the animals and plants of the Hill Country is probably the growth of urban areas along the southeastern side of the Hill Country, along the Balcones Escarpment. San Antonio has more than a million people, Austin is heading toward a million, and the counties along Interstate Highway 35 are among the most rapidly growing in the United States. The population of Hays County grew 60 percent in the decade from 2000 to 2010, adding 60,000 people. Population projections suggest a future with over 5 million people in the Hill Country by the year 2050. "Hill Country Living" is advertised by developers as a way of life combining the amenities of the city with the nature of the Hill Country. However, in many developments the ground is stripped to bedrock before the houses and roads are built. Native vegetation is replaced by turf grass and various introduced landscape plants.

Among the many effects of such development, perhaps of most concern are the increasing water use necessary to sustain the growing population and the desire of many homeowners for green

Suburban development in the Hill Country

lawns during the hot, dry summers. As wells are drilled and more water pumped, the effects will be felt in decline of aquifer levels. As explained earlier, the aquifer level is directly related to spring flow, which is necessary for the continued existence of the endangered species that dwell in the springs. The main problem is the declining aquifer level during a series of drought years. A multiyear drought like that of the 1950s is the most severe threat. The Edwards Aquifer Habitat Conservation Plan has also been developed and approved. The goal of this plan is to prevent the extinction of the aquatic species in the event of a severe drought.

A related concern is the development in the Hill Country occurring directly on the Edwards Aquifer recharge zone, which runs right through the area that is rapidly urbanizing. As water passes over and through vegetated land, pollutants can be filtered out before they reach the aquifer. However, urban runoff from streets, parking lots, and pesticide- or fertilizer-treated lawns is usually much higher in pollutants than water from ranch land and may enter the aquifer without any filtration if the vegetation has been cleared. Currently, water from the aquifer meets the drinking water standards without any treatment. If the aquifer becomes polluted, millions of dollars will have to be spent by the cities using this water to build and operate water treatment plants.

For this reason, since 1997 San Antonio has been purchasing land over the recharge zone and preventing development on those plots through the San Antonio Aquifer Protection Program. In 2000, 2005, and 2010 citizens voted for a one-eighth-cent sales tax with the proceeds to go to land purchases for aquifer protection. Almost 100,000 acres have already been purchased or protected by conservation easements, in which the government purchased the development rights for the land. Whether enough land will be acquired to prevent pollution of the aquifer is not yet known. Efforts to preserve land near Austin will have similar effects, including projects there such as the Balcones Canyonlands National Wildlife Refuge that have focused on protection of habitat for the Golden-cheeked Warbler. Lands protected over the next decade or so are likely to be the only natural habitats remaining in the eastern counties of the Hill Country.

Urban development also impacts the karst-dwelling invertebrates, many of which are found in only one or a few cave systems. As these are paved over, the ecosystems in the caves are cut off from their source of energy, which is plant and animal material from aboveground. In many cases, small areas around the cave entrance are now protected from development.

CHAPTER ONE Trees and Shrubs

Trees and shrubs provide the framework for the ecological communities of the Hill Country. The distribution of other species is often strongly related to tree distribution. For example, the endangered Golden-cheeked Warbler is dependent on Ashe juniper for its nesting materials and thus will not be found nesting in areas lacking this tree. Trees and shrubs often provide large amounts of food for animals in the form of fruits and nuts. The leaves and wood are seldom important food sources because they usually contain chemicals that are toxic to animals or that make it difficult for the plant material to be digested. New growth on the ends of branches is an exception and is often browsed by deer and other animals.

Much of the land in the Hill Country is dominated by Ashe juniper and live oak. However, there are dozens of other species that occur frequently to rarely. The species found here are an interesting mix typical of desert, forest, and plains regions. The described species are arranged alphabetically by scientific family name. Species in the same family are listed alphabetically by genus and species. Leaves, flowers, and fruit are all helpful in identifying specimens.

Woodland of Ashe juniper and Live oak at Garner State Park

Simple leaves, long and grass-like
- Beargrass
- Texas sotol
- Spanish dagger
- Buckley's yucca
- Twist leaf yucca

Simple leaves, narrow
- Black willow
- Bald cypress
- Poverty bush
- Retama (if leaflets have fallen o
- Beebush

Simple leaves, oval and unlobed
- Texas persimmon
- Cedar elm
- Texas madrone
- Elbowbush
- Cenizo
- Lindheimer's silktassel
- Netleaf hackberry
- Chinese tallow
- Cottonwood
- Texas lantana
- Live oak
- Common buttonbush

Simple leaves, lobed
Shin oak
Post oak
Lacey oak
Blackjack oak
Spanish oak
Bigtooth maple
American sycamore
Dwarf palmetto

Once-compound leaves
Box elder
Texas ash
Agarito
Skunkbush
Evergreen sumac
Soapberry
Rattlebox
Texas mountain laurel
Pecan
Mexican buckeye
Texas kidneywood
Retama

Twice-compound leaves
Chinaberry
Honey mesquite

Bigtooth Maple
Acer grandidentatum   Aceraceae   Maple Family

Bigtooth maple is the star attraction in the fall at Lost Maples State Natural Area, where it is abundant along the Sabinal River. The leaves are about 3 inches wide and turn orange or red in November. It is also found in various other canyons in the Hill Country, as well as in the mountains of West Texas and several of the western states. How did the maples get "lost" in the Hill Country? Maples usually occur in areas with moister soil than most of that in present-day Central Texas. During the ice ages, the climate in Texas was much cooler and wetter; hence, it is likely that maples were more widely distributed at that time. As the climate became warmer and drier during the past 10,000 years, the only place that the maples could survive was in shady canyons along spring-fed streams, where there is a steady supply of groundwater.

Boxelder

Acer negundo   Aceraceae   Maple Family

Boxelder is in the same genus as the bigtooth maple but cannot be confused with it because the leaves of boxelder are compound. Usually they have three or five leaflets, but they can have up to nine. The leaves are very similar to those of poison ivy, and poison ivy can appear to come from a trunk when it is growing along the branches of a tree. The leaves of poison ivy are alternate, whereas those of boxelder are opposite. In the fall, the paired, winged seeds easily identify it as a maple. Boxelder occurs widely in North America and Central America.

Skunkbush

Rhus aromatica   Anacardiaceae   Sumac Family

This sumac has three leaflets per leaf, the leaves are alternate, and they are deciduous. It has small yellow flowers in the spring and red fruit in clusters starting in June. The branches and leaves have a distinct odor when broken. Skunkbush is native to the western half of the United States.

Evergreen Sumac

Rhus virens   Anacardiaceae   Sumac Family

This shrub has evergreen, alternate, compound leaves with five to nine leaflets. The leaflets are dark green and shiny, do not have teeth, and are retained in the winter. The small white flowers grow in clusters, and the fruits are red and hairy. Evergreen sumac flowers in the summer after rains, and fruits are produced in the fall. It occurs on rocky hillsides from Central Texas west to New Mexico and Mexico.

Dwarf Palmetto
Sabal minor   Arecaceae   Palm Family

This is one of two native species of palms in Texas. The leaves are fan shaped as in most palms and grow from a subterranean stem. Dwarf palmetto grows as a shrub along some of the Hill Country streams, as well as many wet areas throughout East Texas and the southeastern United States. The Hill Country populations are probably relicts from a broader distribution during the ice ages. The other native palm in Texas is a tree (*S. mexicana*) and occurs naturally only in the Rio Grande Valley and southward into Central America. Dwarf palmetto is common along the creek at Honey Creek State Natural Area in Comal County.

Texas Sotol

Dasylirion texanum   Asparagaceae   Asparagus Family

Texas sotol is common on rocky hillsides of the Hill Country. Two other species of sotol occur in West Texas. The long, thin leaves have curved spines along the edges. The flowers are located on a thick stalk that can be more than 10 feet tall. Native Americans used the bulb at the base of the leaves to make flour. Sotol has been widely adopted as an attractive landscape plant that never needs supplemental water.

Beargrass
Nolina texana and *N. lindheimeriana*
Asparagaceae   Asparagus Family

Beargrass looks like a clump of tall grass, but the leaves are green year-round, and in the spring the plant produces flowering stalks with small white flowers. *Nolina texana* has narrow leaves with smooth margins (or widely spaced teeth). *Nolina lindheimeriana* has saw-toothed leaf margins (another common name is devil's shoestring), and it forms much thinner clumps than *N. texana*.

Buckley's Yucca

Yucca constricta   Asparagaceae   Asparagus Family

This yucca can be recognized by the almost perfectly globe-shaped cluster of leaves. The leaves are about ½ inch wide and up to 2 feet long. The margins have long threads that curl off. The pale flowers are produced on a stalk that extends several feet above the leaves. This species is found only in Texas. Yuccas are pollinated by a specific group of moths called yucca moths. Yuccas also can reproduce vegetatively, sprouting new plants from the base.

Twist Leaf Yucca
Yucca rupicola   Asparagaceae   Asparagus Family

Twist leaf yucca is endemic to the Texas Hill Country and is very common in many parts of the region. The twisted leaves easily distinguish it from other yucca species. It is also usually much smaller than the other common yuccas of the Hill Country and does not form a trunk. Twist leaf yuccas produce a stalk with white flowers in the spring.

Spanish Dagger
Yucca treculeana   Asparagaceae   Asparagus Family

This yucca has broad, spearlike leaves, 1–3 inches wide and up to 3 feet long. The woody stem can grow 10 feet tall. The dead leaves bend back on the stem below the crown of leaves. The white flowers are produced on a stalk just above the leaves in March and April. It occurs from the southern part of the Hill Country through South Texas and northern Mexico.

Poverty Bush, Roosevelt Weed
Baccharis neglecta   Asteraceae   Aster Family

This tall shrub is most often found in disturbed areas such as neglected fields and in dry riverbeds. It has narrow green leaves that are about 1.0–2.5 inches long. In the fall the female plants produce attractive silky plumes on the flowers. It occurs throughout Texas and into Mexico. Poverty bush produces large numbers of seeds and can resprout or grow from seed after fires or cutting.

Agarita
Berberis trifoliolata   Berberidaceae   Barberry Family

Agarita is one of the most common shrubs in the Hill Country, often forming an impenetrable ring around the base of large trees. It is easily recognized by its three leaflets, each of which has several very sharp points. The early-spring flowers are small and yellow and attract a wide variety of insects. The berries are bright red and can be used to make jelly. The bright yellow root has been used as a dye. A close relative, Texas barberry (*B. swaseyi*), is endemic to the Texas Hill Country but is much less common than agarita. It is very similar to agarita but has five to nine leaflets per leaf.

Netleaf Hackberry

Netleaf Hackberry
Celtis reticulata

Sugar Hackberry
Celtis laevigata   Cannabaceae   Hemp Family

Netleaf hackberry is native to the western United States, and sugar hackberry is native to the southeastern United States. Their ranges overlap in the Hill Country. Both species have simple leaves that come to a point. The leaves of netleaf hackberry are rough on the upper surface, while those of sugar hackberry are smooth and the shape is narrower. Both species have small round fruits about ¼ inch in diameter. Both species often have wartlike growths on the bark and are frequently parasitized by mistletoe. Hackberry trees are fast growers and common in urban areas even though they are seldom planted. The berries are an important food source for birds and mammals, and the leaves are often consumed by insects.

Ashe Juniper
Juniperus ashei   Cupressaceae   Cypress Family

Ashe juniper, also known as mountain cedar, is a dominant tree in
the Hill Country. It covers the majority of hillsides and upland areas.
Dense stands are called cedar brakes. This tree is easily recognized
by its small, scaly leaves. Male trees produce huge amounts of pollen
in the winter months, leading to "cedar fever" in those people who
are allergic to the pollen. Female trees produce blue cones that look
like berries in the summer and fall. Ashe juniper is native to the Hill
Country but has greatly increased in abundance due to the reduction
in wildfires. The endangered Golden-cheeked Warbler uses the bark
from large Ashe juniper trees to build its nests. The berries of Ashe
juniper are an important food source for birds and small mammals.

Bald Cypress

Taxodium distichum   Cupressaceae   Cypress Family

Much of the beauty of Hill Country rivers can be attributed to the huge bald cypress trees that line their banks. Currently, the state's largest bald cypress, which has a circumference of 37 feet and is 94 feet tall, is located in Real County in the western part of the Hill Country. Bald cypress is a deciduous conifer with leaves that turn rusty-red in the fall. The green balls at the ends of branches are the female cones. Bald cypress often have "knees," or projections from the roots. However, not all bald cypress trees have knees, especially those along Hill Country rivers, and their function is not yet clear to scientists. The wood of bald cypress is very strong and rot resistant, and many of the huge trees that once lined the rivers of the Hill Country were cut down by early settlers for lumber. The Hill Country is the western edge of the range of bald cypress, which occurs throughout the southeastern United States.

Texas Persimmon
Diospyros texana   Ebenaceae   Ebony Family

This is one of the most common shrubs and small trees in the Hill
Country. A distinctive feature is that the edges of the leaves roll under.
The gray bark is smooth and peels off as the trunk and branches grow.
The fruits, which occur only on female trees, start out green and turn
black when they are ripe. They are edible but contain several large
seeds that make them less palatable to humans. The wood of persimmon is extremely hard and dense.

Texas Madrone

Arbutus xalapensis   Ericaceae   Heath Family

Texas madrone is a beautiful small tree with reddish or pink bark that peels off. The leaves are dark green and leathery, the flowers are white, and the fruits are red. This tree is in the same family as blueberries. Texas madrone is fairly uncommon in the Hill Country but also occurs in the mountains of West Texas and south to Guatemala. Small madrone plants are readily eaten by browsers such as deer, which may play a role in the rarity of the tree in the Hill Country.

Chinese Tallow
Sapium sebiferum   Euphorbiaceae   Spurge Family

Chinese tallow is a nonnative invasive tree that is spreading in the Hill Country. Its leaves look somewhat like those of the native cottonwood but have smooth rather than scalloped margins. The leaves turn red or yellow in the fall. Chinese tallow can spread quickly because of its rapid growth, abundant seeds, and production of suckers (sprouts from the stem or roots). Dense forests of these trees can exclude native species, and their leaf litter in water has been shown to affect tadpole survival. They have become especially abundant on the Gulf coastal plain, where they have spread into native prairies.

Texas Mountain Laurel
Dermatophyllum secundiflorum   Fabaceae   Legume Family

This common shrub is known for its beautiful purple clusters of flowers that smell like grape Kool-Aid. It is widely used in landscaping throughout Central Texas because of its beauty and drought tolerance. However, it is rather slow growing. The evergreen leaves are compound with 5–14 leaflets. The pods contain hard red seeds, which are toxic to both humans and animals.

Texas Kidneywood
Eysenhardtia texana   Fabaceae   Legume Family

The compound leaves of Texas kidneywood are about 3 inches long and have up to 40 leaflets. Crushing the leaves results in a strong odor. The flowers are white and arranged in a spike and are visited by many species of insects. The shrubs have many branches and lack thorns. Deer like to browse this species. It lives in the southern half of Texas and in Mexico.

Retama
Parkinsonia aculeata   Fabaceae   Legume Family

Retama is a tropical species that reaches its northern limit in Central Texas. It has distinctive long leaves with tiny leaflets, which are often shed from the grasslike leaf stalk. The bark is green, and the branches have sharp spines. The yellow flowers are produced in the summer and are followed by pods 2–4 inches long. It usually grows in open, disturbed areas such as pastures or dry streambeds.

Honey Mesquite

Prosopis glandulosa   Fabaceae   Legume Family

The leaves of mesquite have two main leaflets, each of which has many pairs of narrow subleaflets. The tree has an open, airy appearance. The trunks almost always grow at an angle from the ground, and mesquites frequently resprout from the base. Spines are common on the branches at nodes. The flowers are yellow and found in 2-inch spikes at the ends of branches. The pods are slender and green or brown. The dense wood is favored for cooking barbecue and is also used for making furniture. Native Americans and European settlers used mesquite in dozens of other ways as well.

Rattlebox

Sesbania drummondii Fabaceae Legume Family

Rattlebox is a tall shrub found in open, wet areas such as along streams or ponds. The leaves are compound with about 40 leaflets per leaf. The yellow flowers hang in clusters. The fruits are a distinctive pod that has four wings along the sides. When dry, the seeds rattle in the pod. The seeds are poisonous to livestock and humans.

Texas Red Oak, Spanish Oak
Quercus buckleyi   Fagaceae   Beech Family

Texas red oak leaves are deeply lobed with long points at the ends. The leaves turn red in the fall. The acorns of these and other oaks are an important food source for mammals. Texas red oaks are common small- to medium-sized trees in the Hill Country on limestone hillsides. Many of the oak species are susceptible to the fungus that causes oak wilt, which has been killing oak trees throughout Texas and other states. Recent evidence suggests that this fungus was introduced to the United States. Information on protecting oaks from this disease is available on the website of the Texas Forest Service.

Texas Live Oak, Plateau Live Oak
Quercus fusiformis (*Q. virginiana* var. *fusiformis*)
Fagaceae   Beech Family

Live oaks are abundant throughout the Hill Country. The leaves of Texas live oak are usually not lobed and are about 1–3 inches long. Small trees commonly sprout from the roots of large trees, and these sprouts usually have hollylike leaves with pointed lobes. This species is sometimes considered a variety of the southern live oak (*Q. virginiana*), which grows farther east, due to hybridization in Central Texas. Live oaks can be infected by a fungal disease called oak wilt (see red oak). Live oaks commonly have round pinkish or brown galls on them, caused by the mealy oak gall wasp.

Lacey Oak
Quercus laceyi   Fagaceae   Beech Family

This medium-sized oak tree is characterized by leaves that are shallowly lobed (similar to live oak leaves) but have a waxy coating and blue-green color. This uncommon species is found only on the Edwards Plateau and in northeastern Mexico. It grows on rocky limestone hillsides.

Blackjack Oak
Quercus marilandica   Fagaceae   Beech Family

Blackjack oak leaves have pointed tips at the ends of three large lobes. The leaves are not deeply lobed like those of Texas red oak. The trees are fairly small, growing to only about 30–40 feet tall. In the Hill Country, they usually grow in dry, sandy soil such as in the Llano Uplift area. This species occurs throughout the southeastern United States and reaches the western edge of its range in the Hill Country.

Shin Oak, Bastard Oak, White Shin Oak
Quercus sinuata   Fagaceae   Beech Family

The leaves of shin oak are lobed, not pointed, and have sinuous (wavy) margins. The bark is shaggy and light gray. In very dry areas, it often grows as thickets rather than trees. Shin oaks (as well as Texas red oak) are often used by the endangered Black-capped Vireo as nest sites. Its range is from Texas to North Carolina and northern Florida.

Post Oak
Quercus stellata   Fagaceae   Beech Family

The leaves of post oak have round lobes, with the three largest at the end forming a thick cross. Post oaks reach their western boundary in the Hill Country and are much more abundant in the eastern part of Texas. They grow well in sandy and low-nutrient soils. They are often found with blackjack oaks.

Lindheimer's Silktassel
Garrya ovata subsp. *lindheimeri*   Garryaceae   Garrya Family

This plant is unique to the Hill Country and northern Mexico. It is an attractive shrub with oval leaves that are fairly thick and usually not damaged by insects. The flowers hang down from the ends of branches, and the fruit is small dark berries. This plant is often found growing in the shade on limestone bluffs. Plants with fruit are female because the male and female flowers are on different plants. The plant is named for Ferdinand Jacob Lindheimer, who was a prolific collector of Central Texas plants in the mid-1800s and is known as the "Father of Texas Botany."

Pecan
Carya illinoinensis   Juglandaceae   Walnut Family

Pecans have large compound leaves with 11–17 leaflets. They leaf out later in the spring than most other trees. Each green fruit is about 1.5 inches long and splits open in the fall to release a brown nut. Pecan trees occur primarily in the rich soil along rivers and are an important food source for mammals and turkeys. Pecan is the state tree of Texas. A related species in the Hill Country, the Texas walnut (*Juglans microcarpa*), is a small tree that has narrower leaflets than pecan and a round fruit less than ¾ inch in diameter.

Chinaberry
Melia azedarach Meliaceae Mahogany Family

Chinaberry has large leaves, up to 2 feet long, that are twice compound with serrated edges on the leaflets. The flowers have five narrow pink petals surrounding a dark purple tube. The fruits are round and yellow. This fast-growing, nonnative tree is invasive and is now found widely in the Hill Country.

Elbow Bush
Forestiera pubescens Oleaceae Olive Family

Elbow bush is a common understory plant in the Hill Country. The branches tend to droop over, giving the shrub an irregular shape. The leaves are simple and opposite. The branches usually come off at right angles, hence the name elbow bush. The flowers lack petals, and the small blue fruits are in clusters.

Texas Ash

Fraxinus texensis   Oleaceae   Olive Family

Texas ash is a small tree with compound leaves, each of which has five leaflets. Each seed has a winglike extension that aids in wind dispersal. This ash species is unique to Central Texas and southern Oklahoma. There are several closely related species of ashes in Texas that occur at the edges of the Hill Country.

American Sycamore

Platanus occidentalis   Platanaceae   Sycamore Family

Sycamores are found along the rivers of the Hill Country and are widely planted in cities as shade trees. They can easily be recognized by the multicolored smooth bark, which peels off. The leaves look somewhat like those of bigtooth maples but are larger. The seeds are produced in a round ball about 1 inch in diameter. Sycamores are among the largest trees in Texas and can reach heights greater than 100 feet. They are often one of the first species to colonize gravel bars formed by floods in river channels.

Common Buttonbush

Cephalanthus occidentalis   Rubiaceae   Madder Family

This shrub is easily recognized by the ball-like flower clusters and fruits, each about an inch in diameter. The leaves are smooth-edged and come to a point. This species is usually found along the edges of streams or ponds and is important in anchoring riverbanks.

Cottonwood

Populus deltoides   Salicaceae   Willow Family

Cottonwood leaves are somewhat triangular in shape and pointed at the tip, with toothed margins. The petiole of the leaf is 2–3 inches long and flattened, allowing the leaves to wave back and forth in even a mild breeze. Cottonwoods occur along streams and can grow to become huge trees. The name comes from the puff of white hairs that help the seeds blow in the wind. Cottonwoods are fast growing and have lightweight wood. The Spanish name for cottonwood is *alamo*, and a famous Spanish mission in San Antonio was named for the cottonwood trees nearby.

Black Willow

Salix nigra   Salicaceae   Willow Family

Black willow is found along streams, lakes, and ponds in the Hill Country. It has long, narrow leaves with small serrations on the edges. The flowers are grouped in cylindrical clusters called catkins. Black willow grows fast and often has multiple trunks.

Western Soapberry

Sapindus saponaria var. *drummondii*   Sapindaceae   Soapberry Family

Western soapberry is a small tree with compound leaves, about 12 inches long, with 10–19 leaflets. The leaves turn yellow in the fall. The fruits are yellow, somewhat translucent globes about ½ inch in diameter. Western soapberry occurs throughout the southern states and into Mexico, in a variety of habitats. The fruit is poisonous and contains saponin, which creates lather when shaken in water.

Mexican Buckeye
Ungnadia speciosa   Sapindaceae   Soapberry Family

The most distinctive characteristic of Mexican buckeye is the fruit, which has three lobes and is brown when ripe. The fruit splits open to reveal three dark brown shiny seeds, each ½ inch in diameter. The compound leaves are up to 12 inches long, with five to seven leaflets. The plant forms a small multitrunked tree or tall shrub. The flowers are pink and appear in the spring before or at the same time as the leaves.

Cenizo, Purple Sage
Leucophyllum frutescens   Scrophulariaceae   Snapdragon Family

Cenizo is a native shrub of Texas and Mexico that is widely used in landscaping. Its main native range is western and southern Texas. It is drought resistant and evergreen, with attractive, soft, silvery leaves. The light purple flowers appear a few days after rains from spring to fall.

Cedar Elm
Ulmus crassifolia   Ulmaceae   Elm Family

Cedar elm occurs widely throughout the Hill Country. It is easily recognized by the small simple leaves, 1–2 inches long, with serrated edges. The leaf surface is rough, and the leaves turn yellow in
the fall. Sometimes the smaller branches have corky wings. This tree
is drought tolerant and is often planted as a shade tree. Unlike other
elms that flower in the spring, cedar elms flower and produce seeds in
the fall. American elm (*U. americana*) and slippery elm (*U. rubra*) are
much less common in the Hill Country, where they reach the western edge of their ranges. Both have much larger leaves than those of
cedar elms.

Beebush, Whitebrush
Aloysia gratissima   Verbenaceae   Verbena Family

Beebush is a slender shrub with small simple leaves (less than 1 inch long), usually in clusters along the stem. The spikes of small white flowers are produced after rains, and as the name suggests, these attract bees and other insects. Beebush provides excellent protective cover for small animals, including birds, reptiles, and mammals. It occurs on rocky slopes and grasslands in Texas, New Mexico, and Mexico.

Texas Lantana
Lantana urticoides (L. horrida)   Verbenaceae   Verbena Family

Texas lantana is a short, spreading shrub with orange, red, or yellow flowers that grow in clusters. The individual flowers are tubular with four or five lobes. The leaves are rough and have toothed edges. The fruits are small, round, and dark blue or black. Lantana is widely planted due to its colorful flowers and drought resistance. Many cultivars have been introduced to the nursery trade.

CHAPTER TWO

Wildflowers

Wildflowers ARE ONE OF THE MAIN ATTRACTIONS of the Hill Country. The abundance of wildflowers varies from year to year and is primarily dependent on rainfall. There are hundreds of species of flowering plants in the Hill Country, and the most common and recognizable are described here. The wildflowers are arranged by color to facilitate identification. Within each color, they are arranged alphabetically by family name and then by genus and species in each family.

Flowers are composed of four main types of structures, which occur in the same order from the base or outside of the flower to the top or inside. However, not all flowers have all four structures. The outermost structures are the sepals, which usually enclose the flower in the bud stage. The petals are within the sepals and are usually the colorful part of the flower. The stamens are the male parts of the flowers and hold anthers that produce pollen. The carpels are the female parts of the flowers and contain the eggs that are fertilized by pollen.

Flowers of plants in the sunflower family (Asteraceae) are combined to form a flowering head that looks like one flower. The center is composed of many small disk flowers with tiny petals. These are surrounded by ray flowers on the outside, which usually have large showy petals that extend outward.

Indian Blanket, Firewheel
Gaillardia pulchella   Asteraceae   Sunflower Family

Indian blanket is one of the most common and popular wildflowers in Central Texas. The flower head is red in the center and yellow on the edges. The leaves are alternate and mostly unlobed. This species grows easily from seed and is recommended for restoration projects. Peak flowering time is from April to June.

Mexican Hat
Ratibida columnifera   Asteraceae   Sunflower Family

Mexican hat can be recognized by the tall column of tiny disk flowers above the ray flowers. The large petals of the ray flowers at the base are yellow and red or sometimes only yellow. The leaves have many narrow lobes. Mexican hat blooms from spring to fall. This species occurs throughout much of North America.

Cedar Sage

Salvia roemeriana   Lamiaceae   Mint Family

Cedar sage and the closely related tropical sage (*S. coccinea*) have bright red flowers with a tubular shape. Cedar sage has rounded leaves, with the leaves on the upper part of the plant simple but those on the lower part compound with three leaflets. Tropical sage has narrower, pointed leaves. Cedar sage is found primarily in the Hill Country and west to the Big Bend region. Tropical sage is more widely distributed across Texas and the southeastern United States.

Velvet-leaf Mallow

Allowissadula holosericea   Malvaceae   Mallow Family

Velvet-leaf mallow has soft, heart-shaped leaves. It grows to 6 feet tall and is locally common on rocky hillsides on the eastern edge of the Hill Country and in West Texas. The flowers are a pale orange, with five petals and numerous orange stamens. The fruit has five sections, each of which holds two rows of seeds.

Turk's Cap
Malvaviscus drummondii   Malvaceae   Mallow Family

Turk's cap has large, dark green leaves and a bright red flower that looks like a turban with a tube of stamens and pistils projecting upward in the middle. The fruits are also red. This plant is common in the understory of Hill Country woods and is widely used in landscaping because it is tolerant of shade and drought and attracts hummingbirds.

Indian Paintbrush

Castilleja indivisa   Scrophulariaceae   Snapdragon Family

The beautiful red-orange display of the Indian paintbrush is composed of modified leaves (called bracts) that surround the narrow yellow flowers. Occasionally the bracts are white. Prairie paintbrush (*C. purpurea*) is similar but has lobed leaves and bracts of various colors. Both species are partially parasitic on the roots of other plants.

Antelope-horns
Asclepias asperula   Apocynaceae   Dogbane Family

Antelope-horns have a large cluster of yellow-green flowers at the end of the stem in the spring. Each flower has five petals. The plants usually have several stems that bend toward the ground. The leaves are up to 6 inches long and narrow with a pointed tip. This plant is one of the milkweeds (formerly placed in their own family, Asclepiadaceae), which are named for the milky sap that is evident if you break a leaf or stem. This species occurs from Texas north to Nebraska and west to California.

Coreopsis
Coreopsis tinctoria   Asteraceae   Sunflower Family

Coreopsis flower heads are yellow with a dark red-brown center. The leaves have narrow lobes. This species is very common in open fields and along roadsides. Rock coreopsis (*C. wrightii*) is a closely related species that also occurs in the Hill Country. Underneath the flower, this species has longer outer bracts (phyllaries) than inner bracts, whereas the opposite is true for *C. tinctoria*.

Common Sunflower

Helianthus annuus   Asteraceae   Sunflower Family

Common sunflowers have many yellow ray petals around a large center of brown disk flowers. The plants range from one to several feet in height. The sunflower grown for seeds and oil is derived from this species. Maximilian's sunflower (*H. maximiliani*) is also common in the Hill Country but flowers only in the fall. In contrast to the common sunflower, the leaf blade of Maximilian's sunflower has no distinct petiole.

Cowpen Daisy

Verbesina encelioides   Asteraceae   Sunflower Family

This species, also known as golden crownbeard, has yellow ray petals with three lobes at the tips. The flowers are about 2 inches wide. The leaves and stems are hairy, and the petioles of the leaves have wings, especially where they attach to the stems. The plant is usually 1–3 feet tall. Cowpen daisy typically grows in disturbed areas throughout Texas.

Hairy Wedelia

Wedelia acapulcensis var. *hispida*   Asteraceae   Sunflower Family

Hairy wedelia (also known as zexmenia) is one of several common sunflowers in the Hill Country. It grows in a short clump with a woody base. The flowering head has yellow rays and a yellow center. The leaves are opposite, pointed, and covered with short hairs. Hairy wedelia blooms from April through the fall.

Yellow Stonecrop
Sedum nuttallianum   Crassulaceae   Stonecrop Family

Yellow stonecrop is a small succulent, a plant that stores water in its leaves or stems. The leaves are noticeably thicker than those of most other plants. These plants can survive in dry conditions and are common in the Hill Country on rock outcrops such as Enchanted Rock. The small yellow flowers have four to five petals.

Buffalo Gourd, Stinking Gourd
Cucurbita foetidissima   Cucurbitaceae   Gourd Family

Buffalo gourd grows along the ground on long runners that can extend to 30 feet. The leaves are grayish-green, somewhat triangular in shape, and have an unpleasant smell. The flowers are yellow, and the fruit is green with light stripes. This species occurs from the central and southwestern United States to Mexico. Mature fruits are poisonous to humans, but the roots and seeds were used in a variety of ways by Native Americans.

Texas Queen's Delight
Stillingia texana   Euphorbiaceae   Spurge Family

Texas queen's delight is quite easy to recognize even when not flowering. It has many stems coming out from a woody base, forming a rounded bush about 1–2 feet tall. Each stem has many narrow, dark green leaves with small teeth, and the sap is milky. The flowers are small and green on spikes at the ends of the stems. The fruits are green with three lobes and are clustered at the ends of the stems. This species usually grows on calcareous soils.

Lindheimer's Senna

Senna lindheimeriana   Fabaceae   Legume Family

Lindheimer's senna has inch-wide yellow flowers with five petals that are produced in the fall. The compound leaves are several inches long with 10–12 leaflets. The stems and leaves are soft. The fruits are flattened pods about 2 inches long. The dried brown pods often remain on the dead stalk of the plant through the winter. The plant can grow to 3 feet tall.

Common Mullein
Verbascum thapsus   Scrophulariaceae   Figwort Family

Common mullein has soft, gray, fuzzy leaves arranged in a rosette with a tall stalk arising from the middle. The yellow flowers are produced at the top of the stalk. This species is native to Eurasia but is now widespread in North America. Another common name is cowboy's toilet paper because of the soft leaves.

Dayflower
Commelina erecta   Commelinaceae   Spiderwort Family

The dayflower has two blue petals and a smaller, less conspicuous white petal. The leaves have parallel veins and clasp around the stem. Each flower lasts less than one day. Dayflower blooms from May to October in Central Texas. This species occurs in much of the United States as well as Central and South America, Asia, and Africa.

Texas Bluebonnet

Lupinus texensis   Fabaceae   Legume Family

The state flower of Texas is actually any of the six species of blue-bonnets that grow in the state. This species is the one usually seen in the Hill Country. The leaves are palmately compound with five leaf-lets radiating from a central point. The flowers are blue and white. The center starts out white and turns red. The plants are winter annu-als that grow through the winter and flower in the spring. The fruits are small, green, hairy pods.

Mealy Sage
Salvia farinacea   Lamiaceae   Mint Family

At first glance, the blue-purple flowers of mealy sage might be mistaken for a bluebonnet. However, the leaves of mealy sage are simple, not compound, and the stem is square. The flowers are on a stalk extending 12 inches or more above the leaves. Mealy sage is a perennial that blooms in spring to summer and again with fall rains. This species is related to the sage used in cooking.

Wild Petunia

Ruellia spp.   Acanthaceae   Wild Petunia Family

Wild petunia has purple flowers with five petals. The leaves are simple with wavy, toothed margins. Wild petunias are usually in bloom from spring until fall. There are several similar species in the Hill Country in this genus. One species (*R. metziae*) has white flowers.

Texas Thistle
Cirsium texanum   Asteraceae   Sunflower Family

Thistles have many purple disk flowers but no ray flowers in the flower head. The leaves have many prickly teeth and dense hair on the underside. Several thistles in the genus *Cirsium* are native in Texas, but sow-thistles (genus *Sonchus*) were introduced from Europe. Sow-thistles have yellow flowers and are common lawn weeds.

Dotted Blazing Star

Liatris punctata   Asteraceae   Sunflower Family

This blazing star has one or more spikes of purple flowers rising about
1–2 feet above the ground. The leaves are narrow and 2–4 inches long.
They bloom in the fall. The flowers attract butterflies, including the
fall migration of monarchs. This species is widely distributed across
the prairies of North America.

Skeleton Weed

Lygodesmia texana   Asteraceae   Sunflower Family

The flowering heads of skeleton weed look like a single large flower but are actually composed of many flowers. The large pink petals are part of the ray flowers. In the center of the flower head are several tubelike disk flowers. Skeleton weed gets its name from its mostly leafless appearance. It prefers calcareous soils and blooms from spring to fall.

Small Palafoxia

Palafoxia callosa   Asteraceae   Sunflower Family

Small palafoxia has small purple-to-white flowering heads with disk flowers but no ray flowers. The plant is spindly with narrow leaves. It blooms in the summer and fall. Rose palafoxia (*P. rosea*) occurs more rarely in the Hill Country and is very similar to small palafoxia but has longer bracts at the base of the flower.

Spiderwort
Tradescantia spp.   Commelinaceae   Spiderwort Family

Several species of spiderworts occur in the Hill Country, and they are fairly similar in appearance. The flowers have three purple petals of equal size, and the leaves have parallel veins. *Tradescantia edwardsiana*, *T. humilis*, *T. pedicellata*, and *T. subacaulis* are all endemic to Texas. Other species found here have wider distribution in the eastern United States.

Morning Glory
Ipomoea cordatotriloba   Convolvulaceae   Morning Glory Family

This is the most common species of morning glory in the Hill Country. It has purple flowers about 2 inches wide and leaves that vary in shape but often have three pointed lobes. It is a vine that spreads over other plants or on the ground. It blooms from spring through fall.

Mountain Pink

Zeltnera (Centaurium) beyrichii   Gentianaceae   Gentian Family

Mountain pink has four to five small pink petals that join at the base to form a tube. The plant looks like a small bouquet. The leaves are narrow and about an inch long. Mountain pink usually occurs on dry limestone outcrops. Two other species in this genus also occur in the Hill Country, rosita (*Z. calycosa*) and Lady Bird's centaury (*Z. texensis*). Rosita usually has one main stem from the base, and Lady Bird's centaury has tiny flowers, only ¼ inch in diameter.

Horsemint, Beebalm
Monarda citriodora   Lamiaceae   Mint Family

Horsemints have rings of tubelike flowers around the stem, with leafy bracts in between each ring of flowers. The leaves have a lemon scent, and the flowers attract bees and butterflies. The flowers range from white to purple. *Monarda citriodora* is the most common species of horsemint in the Hill Country, but other species occur, especially on the sandy soils of the Llano Uplift.

Winecup

Callirhoe spp.   Malvaceae   Mallow Family

Three species of winecup occur in the Hill Country. They have five
large purple to pink (sometimes white) petals. The flowers start as a
closed cup and gradually open wider. The leaves are lobed. Winecups
flower in the spring to early summer.

Pink Evening Primrose

Oenothera speciosa   Onagraceae   Evening Primrose Family

Pink evening primrose, also known as buttercup, has large pink flowers with four petals and eight yellow stamens. Leaves are alternate
and 1–4 inches long. This species is often very abundant on roadsides
throughout Central Texas.

Drummond's Wood Sorrel
Oxalis drummondii   Oxalidaceae   Wood Sorrel Family

The leaves of wood sorrel are compound with three leaflets and resemble those of a clover. The flowers are bright pink to purple with five petals. This small wildflower blooms in the fall. Drummond's wood sorrel is native to Texas and northern Mexico.

Golden-eye Phlox

Phlox
Phlox drummondii   Polemoniaceae   Phlox Family

Phlox have purple, red, or pink flowers, often with very intense color. There are five petals, which are united at their base to form a short tube. Golden-eye phlox (*P. roemeriana*) has a bright yellow center. Both species flower from March to May.

Agalinis

Agalinis edwardsiana and *A. heterophylla*
Scrophulariaceae   Figwort Family

Agalinis has pink flowers with purple dots and two yellow lines
inside the tube formed by the five petals. The leaves are narrow and
only 1 inch long. The two species in the Hill Country can be separated by the length of the pedicel, the stem that holds the flower. In
A. edwardsiana, which is endemic to the Hill Country, the pedicels are
long, usually about 1 inch. The pedicels of *A. heterophylla* are less than
⅛ inch long.

Silverleaf Nightshade
Solanum elaeagnifolium   Solanaceae   Nightshade Family

This plant can be recognized by the silvery, hairy leaves and light purple, star-shaped flowers. The stamens in the middle of the flower are bright yellow. The leaves are simple and unlobed. The yellow fruits look like small tomatoes hanging down on their stems, but they are toxic. Buffalo bur (*S. rostratum*) is a similar plant with yellow flowers and prickly leaves and fruits.

Prairie Verbena

Glandularia bipinnatifida   Verbenaceae   Verbena Family

This attractive species is common on roadsides and open areas. The bright purple flowers grow in showy clusters. Each flower is tubular with five lobes. The leaves are divided into narrow lobes and are opposite. It is an important nectar plant for insects.

Old Plainsman
Hymenopappus scabiosaeus   Asteraceae   Sunflower Family

Old plainsman has clusters of small white flowers atop a single stem that is 1–3 feet tall. The leaves have long, narrow lobes. Old plainsman is common along roadsides and in fields in the spring. Two other species in this genus also occur in the Hill Country but are less common (*H. artemisiifolius* and *H. tenuifolius*).

Blackfoot Daisy
Melampodium leucanthum   Asteraceae   Sunflower Family

Blackfoot daisy has white ray flowers with yellow disk flowers at the center. The plant forms a small round clump about 12 inches tall. The leaves are narrow and rough. Blackfoot daisy blooms throughout the summer and is common in open areas throughout the southern Great Plains.

Frostweed

Verbesina virginica   Asteraceae   Sunflower Family

Frostweed has a tall single stem, up to 6 feet tall, with four to five wings along its edges. The simple leaves are toothed and alternate. The flowers are white and arranged in a large cluster. It is common along creeks and under trees but can also be found in open fields. The name comes from the remarkable ice sculptures that form from the sap exuded from the stems on the first below-freezing days of winter.

Snow-on-the-mountain

Euphorbia marginata   Euphorbiaceae   Spurge Family

This plant is distinctive in shape and color even from a distance. The single stem is 1–3 feet tall with smooth green leaves, and it is topped by three branches radiating out and covered with white-margined leaves. The flowers are surrounded by five small white bracts that look like petals. The flowers themselves are tiny and clustered onto a structure called the cyathium.

Crow Poison
Nothoscordum bivalve   Liliaceae   Lily Family

This relative of the onion also grows from a bulb, but this species lacks the odor of wild onions. It is common on roadsides, mowed areas, and open fields. The 6–12 small white flowers, each with six white tepals, form a group called an umbel on the end of each stem. The leaves are narrow with parallel veins.

Rain Lily
Zephyranthes chlorosolen and *Z. drummondii*   Liliaceae   Lily Family

Two species of rain lily occur in the Hill Country. Both have white
flowers with three petals and three sepals that appear a few days
after heavy rains. The petals and sepals, called tepals, are similar.
Zephyranthes drummondii has larger flowers, with a tube that is 3–7
inches long. It usually flowers in the spring. *Zephyranthes chlorosolen*
has smaller flowers, with a tube that is 1.0–1.5 inches long. It usually
flowers in the summer or fall.

White Prickly Poppy
Argemone albiflora   Papaveraceae   Poppy Family

White prickly poppy has abundant prickles on the stems and the edges of the leaves and a large showy white flower with yellow center. It is common in fields and other disturbed areas. *Argemone aurantiaca* is another species of prickly poppy that occurs in the Hill Country. The lower surfaces of the leaves have prickly hairs throughout, rather than just on the veins as in *A. albiflora*.

Pigeon-berry

Rivina humilis   Phytolaccaceae   Pokeweed Family

Pigeon-berry is a low-growing plant with a 2-inch-long group of white flowers on the end of the stem. Each flower has four sepals but no petals. The flower buds are pink before opening. After flowering, the stems hold shiny red berries. The leaves are simple and have smooth or wavy edges. Pigeon-berry usually grows in shaded areas. As their name implies, the fruits are often eaten by birds.

Windflower

Anemone berlandieri   Ranunculaceae   Buttercup Family

Windflower is a common small plant in the early spring in lawns and prairies. The flowers are white to purple. The leaves have three parts or lobes. Two-flower anemone (*A. edwardsiana*) is endemic to the Edwards Plateau. It has two or more flowers per plant, while windflower always has only one flower per plant. Desert anemone (*A. tuberosa*) is a similar plant of the southwestern desert that occurs in the western part of the Hill Country.

Old Man's Beard
Clematis drummondii   Ranunculaceae   Buttercup Family

This vine is most distinctive when the fruits have long silky hairs and look like puffs of cotton. The flowers have white sepals but no petals. The leaves are compound with five or seven leaflets. Old man's beard is common on fences and climbing over shrubs.

CHAPTER THREE Cacti, Vines, Grasses, Ferns, and Other Plants

T HIS CHAPTER DESCRIBES some of the common plants that are not trees, shrubs, or wildflowers. The cacti, which represent a single plant family (Cactaceae), are conspicuous because of their distinctive fleshy stems and sharp spines. Spines are thought to be modified leaves and thus are different from thorns, which are modified branches. Many of the cacti have beautiful flowers that are typically present for only a short time in the spring. Vines do not represent a single family or taxonomic group but instead are a growth form, much like the terms "tree" and "shrub." Some vines simply grow around or on top of structures such as other plants, while others have small tendrils that twist around branches or fence wires. Grasses are members of the family Poaceae within the flowering plants. A wide variety of grasses grow in the Hill Country, both native and introduced. A few distinctive species are described here. See the book *Grasses of the Texas Hill Country* by Brian Loflin and Shirley Loflin for a more complete source. Ferns are a group of plants that are quite different from most of the other plants in this book. They lack both seeds and flowers. Their spores are typically produced in structures on the undersides of the fronds (fern leaves). The other plants in this section represent other groups of seedless plants such as mosses and liverworts. These represent separate lineages from the ferns and seed plants. In each section of the chapter, the plants are arranged alphabetically by family and then genus and species.

Pineapple Cactus, Grooved Nipple Cactus
Coryphantha sulcata   Cactaceae   Cactus Family

The stems of pineapple cactus are about 2–3 inches in diameter and length. The spines of this cactus are located on tubercles that spiral around the plant. Each tubercle usually has six to eight spines radiating from its center, and one to three spines extending out. The flowers appear in the spring and are yellow with pink at the base. This species is found from north-central Texas to northern Mexico.

Tasajillo, Pencil Cactus
Cylindropuntia (Opuntia) leptocaulis   Cactaceae   Cactus Family

Tasajillo has narrow, cylindrical stems that branch to form variously shaped bushes with many spines. The flowers are yellowish-green, and the fruits are red. This plant is very common along fence lines and can reach 3–5 feet in height. It occurs from Texas to Arizona and south to central Mexico.

Horse Crippler
Echinocactus texensis   Cactaceae   Cactus Family

Horse crippler can be recognized by the stout, curved spines and ribbed hemispherical stem. Horse crippler is native to Texas, New Mexico, and northern Mexico. The name suggests the danger of stepping on the stout spines. During drought the stem is often mostly below ground level, and after rains it can swell up to about 2 inches aboveground. The flowers are usually pink, and the fruits are red.

Claret-cup Cactus
Echinocereus coccineus   Cactaceae   Cactus Family

The claret-cup cactus has striking bright red flowers and grows in clumps on rocky hillsides. These cacti have four to six straight radial spines but often lack central spines. Flowering occurs from March to July. This species occurs from the Hill Country to West Texas as well as in Arizona and New Mexico.

Lace Cactus
Echinocereus reichenbachii   Cactaceae   Cactus Family

This is one of the most common cacti in the Hill Country. It grows as a cylinder only a few inches tall. The cylinder has several to many ribs that are covered in radiating spines, with the spines so dense that they obscure the stem. The flowers are pink or pale purple. Lace cactus is usually found in rocky areas of limestone or granite. While these cacti are beautiful and look like they would make nice plants for a windowsill or cactus garden, they are slow growing, so please leave them in their wild surroundings where they can be enjoyed by others. It is illegal to remove plants from public lands.

Texas Prickly Pear
Opuntia engelmannii var. *lindheimeri*   Cactaceae   Cactus Family

Texas prickly pear is an abundant plant throughout the Hill Country and can grow into a shrub several feet tall. It has large green pads with cream to yellow spines. The flowers are yellow, and the fruits are red or purple. The plants occasionally grow on top of large branches of trees such as live oaks.

Poison Ivy

Toxicodendron radicans   Anacardiaceae   Sumac Family

Poison ivy has three shiny leaflets per leaf. It often grows up trees and has hairy aerial roots that anchor it to the bark. It is usually found at the edges of woods where there is adequate sunlight. The leaves and stems contain an oil that causes itching and blistering if it contacts your skin, although some people are less affected by it. The leaves often turn red in the fall, and the berries turn white when ripe.

Greenbrier
Smilax bona-nox   Smilacaceae   Smilax Family

This vine has sharp prickles that easily snag clothing or skin. The leaves and stems are smooth and green and have a firm texture. The leaf shape is variable but never compound. Small tendrils are located at the leaf base. The fruit is black when ripe and held in small clusters. A dense growth of greenbrier is impossible to move through without a knife or clippers.

Cow Itch, Sorrelvine
Cissus incisa   Vitaceae   Grape Family

The leaves of cow itch are thick and fleshy and composed of three leaflets. They have coiled tendrils that attach to the tree or fence they are climbing. The small green flowers appear in June and July. Fruits are small berries that turn from green to black when ripe.

Virginia Creeper

Parthenocissus quinquefolia   Vitaceae   Grape Family

This vine looks somewhat like poison ivy but has five leaflets per leaf. The leaves turn red in the fall. The fruits are small blue berries that grow in loose clusters. Sevenleaf creeper (*P. heptaphylla*) is a related species that is endemic to Central Texas. The best way to distinguish it from Virginia creeper is that its leaflets are fleshier than those of Virginia creeper. The number of leaflets can vary in both species.

Mustang Grape

Vitis mustangensis   Vitaceae   Grape Family

Mustang grape has lobed or unlobed leaves with white fuzzy undersides. The vines are usually found climbing over other plants or fences. The grapes are dark purple when ripe and are edible. The plant can grow very large and develop a thick, woody trunk. Other native grape species in the Hill Country are mountain grape (*V. monticola*), summer grape (*V. cinerea*), and sand grape (*V. rupestris*).

Bushy Bluestem
Andropogon glomeratus   Poaceae   Grass Family

Bushy bluestem is a distinctive tall grass (3–4 feet) with many tufted seeds present in the fall. It forms clumps in wet areas such as along streams. The seeds are eaten by birds, and the grass clumps provide cover for birds and mammals.

Giant Reed

Arundo donax   Poaceae   Grass Family

Giant reed is easily recognized by its tremendous height for a grass, growing up to 18 feet tall. It has long, narrow leaves on unbranched stems. The flowering heads are 1–2 feet long at the top of the stems. Giant reed is native to the Mediterranean region and is used to make the reeds for wind instruments such as clarinets and saxophones. It occurs in the Hill Country in ditches and along rivers.

King Ranch (KR) Bluestem
Bothriochloa ischaemum var. *songarica*   Poaceae   Grass Family

This grass, which is a native of Eurasia, has been widely introduced in Texas as a forage grass for cattle, and it is now abundant on ranches, parks, and preserves throughout the Hill Country. Often large areas are covered by near monocultures of this grass. The stems and leaves are light green but turn straw-colored when mature. The small flowers and seeds are on two to eight branches. Evidence suggests that areas dominated by KR bluestem have lower diversity of native plants.

River Oats, Broadleaf Woodoats
Chasmanthium latifolium   Poaceae   Grass Family

River oats has clusters of large flowering heads that cause the stem to droop. This grass is common on the floodplains along rivers in the Hill Country. It can be used in landscaping for shaded, moist areas.

Mexican Fern
Anemia mexicana   Anemiaceae   Anemia Family

This fern has glossy fronds with four to six pairs of pinnae (leaflets). The spores are produced on two stalks that extend above the fronds. Mexican fern occurs mostly in the southwestern part of the Hill Country. It is the most common fern on the limestone hillsides at Garner State Park.

Southern Maidenhair Fern
Adiantum capillus-veneris   Pteridaceae   Maidenhair Fern Family

The rachis (main stem of the frond) of this fern is a black zigzag, and the pinnae (leaflets) have flaps that fold over the end and cover the sites of spore production. This fern is common on the walls of moist canyons in the Hill Country.

Fairy Sword Fern

Lindheimer's Lip Fern, Fairy Sword Fern
Cheilanthes lindheimeri

Woolly Lip Fern
Cheilanthes tomentosa   Pteridaceae   Maidenhair Fern Family

Lip ferns and fairy swords are well adapted to arid conditions typical of the Hill Country and can be found growing in shaded, rocky areas. These two species are the most common in the Hill Country. Fairy swords are abundant on Enchanted Rock at the bases of boulders. The "lip" of these ferns is a part of the leaf that folds under and covers the structures where spores are formed. The fronds of lip ferns will dry up and die during long periods of drought, but the roots remain alive and will produce new fronds when wet conditions return.

Ovate-leaf Cliffbrake

Pellaea ovata   Pteridaceae   Maidenhair Fern Family

Cliffbrake ferns are another type of fern that occurs on rock outcrops
in the Hill Country. *Pellaea ovata* has a tan zigzag rachis (stem of the
frond) with small oval pinnae (leaflets). *Pellaea atropurpurea* and
P. wrightiana are related species in which the leaf has a dark rachis
that is not zigzag. *Pellaea wrightiana* is common on granite outcrops
and has a small pointed tip at the ends of the pinnae.

Ovate Maiden Fern, Ovate Marsh Fern

Thelypteris ovata var. *lindheimeri*
Thelypteridaceae   Marsh Fern Family

This fern is common on rock walls along small streams and springs
throughout the Hill Country. The leaf is widest in the middle and
comes to a sharp point at the end. Each pinna (leaflet) has indenta-
tions along the edge.

Other Plants

Ball Moss
Tillandsia recurvata   Bromeliaceae   Pineapple Family

Ball moss is far more common in the Hill Country than its relative
Spanish moss. It is easily recognized as a tight clump of grayish stems
and thin leaves. Brown flowering stalks often extend from the clump.
Like Spanish moss, ball moss is not a parasite and usually does not
harm the tree on which it grows.

Spanish Moss

Tillandsia usneoides   Bromeliaceae   Pineapple Family

Spanish moss hangs from the branches of large old trees in the Hill Country, primarily along the southern and eastern edge of this region. It is light grayish-green and can hang down a few feet from its attachment to the tree. It is an epiphyte, not a parasite, because it does not penetrate into the tree's water-carrying tissues.

Smooth Horsetail
Equisetum laevigatum   Equisetaceae   Horsetail Family

Smooth horsetail is a unique plant that reproduces by spores, as ferns do. Plants of this species in the Hill Country are usually unbranched and tall, with green cylindrical stems. Along the stems are circular joints called nodes. When reproductive, smooth horsetails have a spore-producing structure that looks like a small cone at the top of the stem. Smooth horsetails are usually found along rivers and creeks.

Spikemoss
Selaginella spp.   Selaginellaceae   Spikemoss Family

Spikemosses look somewhat like moss but are actually vascular plants. True mosses lack the tubelike tissues that enable vascular plants to transport water up to their leaves. Sand spikemoss (*S. corallina*) occurs on rocky hillsides, often at the base of boulders. This spikemoss usually looks like a small bush, just a few inches tall, with tiny pointed leaves held closely to the stem. Two other species of spikemoss occur in the Hill Country: *S. apoda*, which is found in wet areas and has thin delicate stems and leaves; and *S. peruviana*, which lies prostrate on the ground, forming a mat, and occurs on the granite outcrops of the Llano Uplift.

Mosses
Phylum Bryophyta

While the Hill Country might not seem suitable habitat for moisture-loving mosses, any place with consistent moisture provides habitat for mosses. Look for them along springs and streams and at the bases of trees and rocks. Mosses are difficult to identify to species due to microscopic differences among them. Several plants such as ball moss, Spanish moss, spikemoss, and even algae are often confused with the true mosses (Bryophyta).

Stonewort
Chara spp.   Phylum Charophyta

Stoneworts are a type of algae that are the closest living relatives to the land plants. They are fully aquatic and are common in the streams of the Hill Country. They have stems with whorls of branches, but they lack leaves and roots. The name stonewort comes from the calcium carbonate coating that often forms on the plant.

Liverworts
Phylum Marchantiophyta

Liverworts are similar to mosses in that they do not have the vascular tissue that most plants have. Liverworts are found in wet habitats, such as along streams or near waterfalls, but are much rarer than mosses in the Hill Country. They grow as a series of flat lobes spreading across the ground.

CHAPTER FOUR Fungi and Lichens

FUNGI ARE A GROUP OF ORGANISMS (kingdom Fungi) quite different from plants and animals but actually more closely related to animals than to plants. There are probably hundreds to thousands of species present in the Hill Country, but no list of the species presently exists due to limited research. Fungi typically grow within their food in the form of a network of branched filaments, called a mycelium. The mycelium spreads through the soil, wood, or other dead organic matter as the fungus digests it. Some fungi are parasitic and live on or within living plants or animals. Some fungi are mutualistic, living in symbiosis with plant roots or algae (as in the lichens described in this chapter). Mushrooms are the reproductive structures of many species of fungi. Because the soil is fairly dry in the Hill Country, mushrooms are usually produced after rains when the abundance of moisture allows more rapid growth by the mycelium. Most of the mushrooms are members of phylum Basidiomycota because their spores are produced in microscopic club-shaped structures called basidia. While mushrooms are only apparent when their reproductive structures appear, lichens are highly visible throughout the year. Lichens are grouped here by growth form.

Shelf Fungi
Family Polyporaceae

Shelf fungi are common decomposers of dead wood. Their shelflike reproductive structures project from the wood where the mycelium of the fungus is consuming the wood. The underside of the shelf contains pores, and the spores are produced within these, dropping out of the pores when mature. The shelf structure is usually leathery or woody. Shelf fungi are members of phylum Basidiomycota.

White Juniper Fungus
Family Stictidaceae

The bark of Ashe juniper trees in the Hill Country is usually blotched with white crustose, lichenlike patches. These are not lichens but a type of ascomycete fungi known as *Robergea albicedrae*. According to the scientific report published in 1910 that first described the fungus, it does sometimes kill branches of the juniper trees. The fungus has small gray nodules where the spores are produced.

Lichens

Lichens are widespread, common organisms that are easily overlooked but interesting and beautiful once one takes the time to observe them more closely. They are a mutualistic partnership between fungi and algae in which both species benefit from the association. The lichens are named based on the fungal partner, which is unique for each species of lichen. There are only about 40 known species of algae that participate in lichens, and the majority of lichens use only about a dozen of the most common algae. The algae are members of the green algae or the cyanobacteria. Because the algae are photosynthetic, lichens are producers that get their energy from sunlight. The algae provide some of this energy to the fungi in the form of sugars. The fungus provides protection and attachment to a surface for the algae.

CRUSTOSE LICHENS

Crustose lichens are attached to the surface on which they are growing at all points such that they cannot be lifted off the surface at all. They are common on rocks and the bark of woody plants. Many are brightly colored, and the exposed granite rocks in the Llano Uplift, such as Enchanted Rock, are often covered with many species of various colors. Lichens are less common on exposed limestone because it erodes more quickly.

FOLIOSE LICHENS

Foliose lichens have a leafy appearance and have a definite top and bottom surface but are not completely attached to the surface on which they live. In the Hill Country they are most common on bark but can also occur on rocks. There are probably dozens of species, and the following genera of ruffle and rosette lichens are among the more common.

Ruffle Lichens
PARMOTREMA spp.

Ruffle lichens have broad lobes with ruffled edges. They are loosely attached on tree branches. They are usually gray to pale green, and the lower surface is dark brown to black.

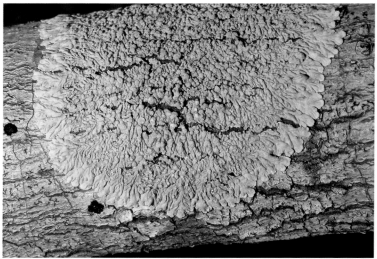
Rosette Lichen

Rosette and Other Similar Lichens

There are several genera in the Hill Country of fairly similar foliose lichens, including *Physcia, Dirinaria, Phaeophyscia,* and *Punctelia.* Rosette lichens range from pale gray to greenish-gray and frequently have spots. The lower surface is usually pale. They are more completely attached to surfaces than are the ruffle lichens.

FRUTICOSE LICHENS

Fruticose lichens have a stalk or bush form, with only a small attachment to a surface and no distinct upper or lower surface of the lichen body. Some grow upright and look like tiny shrubs, while others hang down from the point of attachment. Three distinctive types of fruticose lichens are described here.

Palmetto Lichen

Ramalina Lichens
Ramalina spp.

Several species of lichens in this genus are common and obvious in the Hill Country. They are all grayish-green in color. The palmetto lichen (*R. celastri*) is a fairly large lichen with many branches coming out from the center, looking somewhat like a palmetto leaf. Small, white, spore-producing structures dot the surfaces. Bumpy ramalina (*R. complanata*) has a more bushy form with many small white tubercles and large spore-producing structures at the ends of the branches. *Ramalina peruviana* has narrow branches without tubercles and forms a small shrub or is somewhat hanging.

Orange Bush Lichens
Teloschistes spp.

Orange bush lichens are bright orange clumps that look like tiny shrubs. They usually grow on the branches of trees. There are two species in the Hill Country. *Teloschistes chrysophthalmus* grows in clumps up to 1 inch high and has apothecia (disklike, spore-producing structures) with hairlike projections around the edges. *Teloschistes exilis* has clumps up to 3 inches high and has apothecia without hairlike projections.

Beard Lichens
Usnea spp.

Beard lichens are light green and shrubby or hang down from the point of attachment. The body is composed of one or more central cords with branches extending from them. The sundew beard lichen (*U. cirrosa*) forms a small, dense bush. Each branch ends in a large, round, spore-producing structure with hairlike projections around it. Bony beard lichen (*U. trichodea*) hangs off tree branches, growing up to 12 inches in length. Its central cord has smaller branches extending from it, and the branches are smooth with circular white joints around them.

CHAPTER FIVE   Birds

BIRDS ARE IMPORTANT MEMBERS of the ecological communities of the Hill Country, they are easily observed, and their distribution and abundance are well documented. More than 400 species of resident and migrant birds have been observed on the Edwards Plateau. By the rules of the International Ornithologists' Union, the common names of birds are capitalized, unlike other species. Birds are presented in this chapter in the taxonomic order established by the American Ornithologists' Union.

Two species in particular are strongly associated with the Hill Country, although both also occur outside its boundaries. The Golden-cheeked Warbler and the Black-capped Vireo both breed in the Hill Country during the summer and migrate south to Mexico and Central America for the winter, and both are listed as endangered species by the US Fish and Wildlife Service. The Golden-cheeked Warbler is a small songbird that nests only in mature stands of Ashe juniper and deciduous trees. The Black-capped Vireo occurs in habitats of open shrubland where there is dense shrub vegetation near ground level. Both species are usually detected by listening for their calls in suitable habitat. Because of their endangered status, several areas have been preserved to protect these species, including the Balcones Canyonlands National Wildlife Refuge. However, both species can be observed in many places in the Hill Country.

Because of the high diversity of birds in this region, only the most abundant, likely to be heard or seen, and unique birds are included here. A wide variety of waterbirds such as ducks occur on the lakes and rivers of the Hill Country, especially in the spring, winter, and fall. This assemblage of birds is similar to what you will find on other water bodies in Texas, and therefore most are not included here. Central Texas is also part of the Central Flyway, through which birds from the northern states migrate north in the spring and south in the fall.

(Photograph by Dustin Wyatt)

Wild Turkey

Meleagris gallopavo Order Galliformes, Family Phasianidae

Wild Turkeys are native birds in the Hill Country, where their populations have declined over the past 40 years, based on surveys by Texas Parks and Wildlife Department, but the causes of this decline are unclear. Turkeys nest on the ground, and eggs are often eaten by foxes and raccoons, as well as other predators. Wild Turkeys roost in trees, usually along rivers. A large roosting area is protected in South Llano River State Park near Junction.

Northern Bobwhite
Colinus virginianus   Order Galliformes, Family Odontophoridae

The Northern Bobwhite is a small ground-dwelling quail with a white eyebrow and throat. The call sounds like the name, a short *bob* followed by a whistled ascending *white*. Bobwhites are popular game birds but require large areas with grasslands and brush. Biologists have documented a decline in bobwhites in the Hill Country, which is believed to be the result of overgrazing, increasing shrub cover, and habitat fragmentation.

Great Blue Heron

Ardea herodias   Order Pelecaniformes, Family Ardeidae

The Great Blue Heron is a large wading bird that is common along
the rivers and lakes of the Hill Country. It stands about 4 feet tall,
with a wingspan of 6 feet. These herons are most commonly seen
stalking their prey (fish and frogs) in shallow water and flying over the
water with distinctive slow wing beats. The body is mostly gray with
a black stripe near the top of the head. They are usually silent but
sometimes make a croaking sound when alarmed and honking noises
when flying.

Cattle Egret

Bubulcus ibis   Order Pelecaniformes, Family Ardeidae

Unlike most herons and egrets, Cattle Egrets are usually seen away from water, in pastures where they follow cattle and eat the insects that are disturbed by the cattle. Cattle Egrets are white with a yellow bill. During the breeding season they have orange-yellow feathers on the head and breast. Nesting sites are in tall trees, often near rivers.

Green Heron

Butorides virescens   Order Pelecaniformes, Family Ardeidae

The Green Heron is a small heron that is mostly chestnut, black, and white with greenish wings. The legs are orange. These herons are commonly seen in summer along the rivers and ponds of the Hill Country. They stalk small fish, frogs, and insects in shallow water.

Black Vulture

Coragyps atratus　Order Accipitriformes, Family Cathartidae

Black Vultures are often seen at roadkill along with Turkey Vultures, although they are less common. They have a black head, short tail, and, when seen soaring, white patches at the tips of the wings. They are usually found in small groups. Vultures are sometimes inaccurately called buzzards. Neither species usually makes sounds.

Turkey Vulture

Cathartes aura   Order Accipitriformes, Family Cathartidae

Turkey Vultures are one of the most commonly seen birds in the Hill Country, both along roads where animals have been hit by cars and soaring overhead. The head of adult birds is red, but gray in juveniles. The underside of the wings is gray toward the back and black toward the front. When soaring, Turkey Vultures hold the wings in a slight V shape rather than straight across, as most hawks and eagles do.

(Photograph by H.M. and W.C. Meddaugh)

Red-tailed Hawk

Buteo jamaicensis   Order Accipitriformes, Family Accipitridae

Red-tailed Hawks are common throughout most of North America and are frequently seen in the Hill Country on telephone poles or soaring overhead. They are large hawks with brown above and brown streaks on a white underside. The tail is reddish above and fairly short and wide. Adults are variable in color from darker to lighter phases or "morphs." Juveniles have narrow brown bars on the tail. The call is a distinctive descending scream, *kee-eear*, often heard in western movies. The other common hawk of the Hill Country is the Red-shouldered Hawk (*B. lineatus*), which occurs primarily in forests along rivers. It has reddish and white barring on the chest. The call of the Red-shouldered Hawk is a repeated *kee-ar, kee-ar, kee-ar*.

Killdeer

Charadrius vociferus   Order Charadriiformes, Family Charadriidae

The Killdeer is a medium-sized shorebird but can be found on any type of open ground, not just near water. It has two distinct black bands on its white chest. As the scientific name suggests, these birds make frequent and loud noises. The call is a descending *kill-deee* or a rapid *dee-dee-dee-dee-dee*. The call is made when the bird is disturbed, and when nesting, Killdeer will also try to lure intruders away from their nest, which is on the ground, with a flopping motion that makes it appear that their wing is broken. Killdeer occur throughout most of North America and are resident year-round in Texas.

Ring-billed Gull

Larus delawarensis   Order Charadriiformes, Family Laridae

This gull is the most likely gull to be seen near the lakes of the Hill Country during the fall, winter, or spring. It is white with a gray back, black tail with white spots, and a black band near the tip of the bill. Franklin's Gulls, Laughing Gulls, and Herring Gulls are also sometimes present on the Hill Country lakes.

Rock Pigeon

Columba livia   Order Columbiformes, Family Columbidae

This is the familiar pigeon of cities throughout North America, and it is common in the cities in the Hill Country. Rock Pigeons (also called Rock Doves) are large birds that are native to Europe. Their color varies substantially, but most are gray with a dark head and dark bars on the wings. The call is a soft rolling series of *coo-coo-coo*.

White-winged Dove
Zenaida asiatica   Order Columbiformes, Family Columbidae

These large doves are easily recognized by the white along the edge of the wings when perched. Adults have a light blue ring around each eye. White-winged Doves were once common in Texas only in the lower Rio Grande Valley but now have spread statewide and become one of our most easily sighted birds, especially in urban areas. The call is a cooing *who-cooks-for-you.*

Mourning Dove
Zenaida macroura　Order Columbiformes, Family Columbidae

Mourning Doves have a long, pointed tail and black spots on the wings. The males have pink on the chest and a pale blue crown. The call is a slow and mournful *coo-ooooh-woooo-wooo*, with the second note the highest pitch. Mourning Doves and White-winged Doves are important game birds in Texas.

Inca Dove
Columbina inca　Order Columbiformes, Family Columbidae

Inca Doves are small, light-gray doves with dark-tipped feathers that give an appearance of scales. The underside of the wings has rust-colored feathers that are evident in flight. Its call is a short cooing sound, *coo-coo*, rapidly repeated. Inca Doves are often see in urban areas.

(Photograph by H.M. and W.C. Meddaugh)

Yellow-billed Cuckoo

Coccyzus americanus   Order Cuculiformes, Family Cuculidae

The Yellow-billed Cuckoo is a summer resident in the Hill Country, where it is easier heard than seen in its forest habitats. The call is a loud repeated *ka* that slows down toward the end. A second type of call is a descending cooing, also repeated. The bird is about a foot long, including the long, narrow tail. The wings are rust-colored, the breast is white, and the back is grayish-tan.

Greater Roadrunner

Geococcyx californianus   Order Cuculiformes, Family Cuculidae

Roadrunners are large cuckoos that seldom fly, instead chasing down their insect and small vertebrate prey. They have black-and-white streaks down the body and a long tail. Roadrunners can reach speeds up to 18 miles per hour. They occur throughout open scrubland areas of the southwestern United States and are permanent residents of the Hill Country. The call is a repeated, downward-slurred *coo*.

(Photograph by H.M. and W.C. Meddaugh)

Barred Owl

Owls
Order Strigiformes, Family Strigidae

Barred Owls (*Strix varia*) have a distinctive hooting that sounds like *who cooks for you—who cooks for you all*. They are large owls that lack feather tufts on the head. Their body is mottled with brown and white. Great Horned Owls (*Bubo virginianus*) are large owls that make a low-pitched hooting sound. The "horns" in its name refer to tufts of feathers on the head. Eastern Screech-Owls (*Megascops asio*) are fairly common small owls that are sometimes hard to see but are unmistakable when heard. Their call is a descending whinny. They also have ear tufts, and their color ranges from gray to red.

(Photograph by H.M. and W.C. Meddaugh)

Common Nighthawk

Chordeiles minor   Order Caprimulgiformes, Family Caprimulgidae

Look for Common Nighthawks in the summer, flying over open fields or water bodies, where they catch insects in flight. The most distinctive feature is the white band across the wing, about three-quarters of the way to the tip. The wings are long and pointed. Common Nighthawks spend the winter months in South America.

(Photograph by Tripp Davenport)

Chuck-will's-widow
Antrostomus carolinensis
Order Caprimulgiformes, Family Caprimulgidae

This is another bird that is seldom seen but fairly common and can be heard calling on summer evenings in the Hill Country. The call is just like the bird's name, often repeated again and again. The bird itself is about 12 inches long with brown feathers that provide camouflage in its woodland habitats, where the female lays her eggs in leaf litter on the ground.

Chimney Swift
Chaetura pelagica   Order Apodiformes, Family Apodidae

Chimney Swifts are small, gray, cylindrical birds that are seen in flight and can often be heard chippering in the air before being spotted. They do not perch on wires like the swallows. Because of recent changes in chimney construction and loss of hollow trees, this species is declining. Artificial "chimney" towers have been installed in many parks in this region to provide nest sites for this species. Chimney Swifts migrate to South America for the winter.

(Photograph by H.M. and W.C. Meddaugh)

Black-chinned Hummingbird
Archilochus alexandri   Order Apodiformes, Family Trochilidae

Black-chinned Hummingbirds are tiny, 3–4 inches in length. The males have a black chin with iridescent purple at the bottom, but the purple is often hard to see. Females have a pale throat. The back is green, and the breast is gray and white. This species is the most common hummingbird in the Hill Country and the only one that is a summer breeding resident. The Ruby-Throated Hummingbird (*A. colubris*) migrates through the area, and several other species have been seen occasionally. Listen for the humming sound of their wings as they fly and hover while feeding on the nectar of flowers.

(Photograph by H.M. and W.C. Meddaugh)

Belted Kingfisher
Ceryle alcyon   Order Coraciiformes, Family Alcedinidae

A loud, rattling call and rapid flight over a river are good identification cues for the Belted Kingfisher. This bird has a blue head with a crest and a white band around the throat. Males have a blue band across the chest, and females also have a rust-colored band below that. The head is large with a long, stout beak. These birds dive into water to catch fish. The Green Kingfisher (*C. americana*) is a rarer, smaller kingfisher that is also found in the Hill Country. It has a dark green head and back, white throat, and a rust-colored chest in males. The Ringed Kingfisher (*Megaceryle torquata*) is larger than the Belted Kingfisher and has rust-colored underparts.

(Photograph by H.M. and W.C. Meddaugh)

Golden-fronted Woodpecker

Melanerpes aurifrons   Order Piciformes, Family Picidae

This large woodpecker has a black-and-white laddered back, gold on the back of the neck and above the bill, and white rump. Males have a small red cap. Golden-fronted Woodpeckers range from southern Oklahoma to Nicaragua, primarily in dry woodlands such as those of the Hill Country. Their call is a loud, rolling *churr*.

Ladder-backed Woodpecker

Picoides scalaris   Order Piciformes, Family Picidae

This woodpecker is smaller than the Golden-fronted Woodpecker and has a black-and-white head. The males have a red cap. The call is a short *peek* or a slightly descending whinny. They feed on insects, with the male and female usually foraging together.

(Photograph by Dustin Wyatt)

Eastern Phoebe

Sayornis phoebe   Order Passeriformes, Family Tyrannidae

Eastern phoebes are small gray birds in the flycatcher family. They perch on branches while waiting for insects to fly by, then fly out and catch them. The most distinctive features are the call, which is a whistled *fee-bee*, and their habit of pumping their tail while perched. Some individuals in the Hill Country are summer breeding residents, while others are winter residents that breed farther north.

(Photograph by H.M. and W.C. Meddaugh)

Vermilion Flycatcher

Pyrocephalus rubinus   Order Passeriformes, Family Tyrannidae

Males are bright red on the head and belly, with black on the back and behind the eyes. Females are mostly gray with a streaked breast. This species wags its tail as the Eastern Phoebe does and can also be seen perched on low branches, waiting to fly out and catch insects. Vermilion Flycatchers are near their northern range limit in the Hill Country. They usually migrate south during the winter but have been known to stay here during mild winters.

Western Kingbird
Tyrannus verticalis   Order Passeriformes, Family Tyrannidae

The Western Kingbird is a large flycatcher with a gray head and bright yellow belly. The tail is black with white edges. Like other flycatchers, they are usually seen perched or flying out from their perch to catch insects. They inhabit open areas throughout the western half of the United States. The call is a short *kip* or a rapid chatter.

(Photograph by Dustin Wyatt)

Scissor-tailed Flycatcher

Tyrannus forficatus   Order Passeriformes, Family Tyrannidae

This striking bird has an extremely long tail for its size. The body is light gray with pink on the lower sides and under the wings. The Hill Country is in the middle of its range, which covers the southern central plains from Kansas to Mexico. The call is a series of squeaks that speeds up and rises in pitch over a few seconds, usually repeated many times just before dawn.

Black-capped Vireo

Vireo atricapilla   Order Passeriformes, Family Vireonidae

This small vireo has a black hood with white around the eyes. The back is green, and sides are yellow. Females have a gray hood rather than black. This species is found throughout the Hill Country, as well as in scattered locations north to Oklahoma and south into Mexico, and is listed on the federal endangered species list. Its decline is thought to be due to habitat loss and nest parasitism by the Brown-headed Cowbird, which lays its eggs in other birds' nests. This vireo favors open shrubland with dense, low vegetation. The call is squeaking notes and trills, often described as emphatic.

(Photograph by H.M. and W.C. Meddaugh)

Blue Jay

Cyanocitta cristata Order Passeriformes, Family Corvidae

Blue Jays are fairly common in the eastern half of the Hill Country, especially in cities. They have a blue crest, a black necklace, and a white bar across the wing. The typical calls are a clear, loud *jeer*, or a *wheedle*, usually repeated, but they can also mimic other birds.

Western Scrub-Jay

Aphelocoma californica   Order Passeriformes, Family Corvidae

Western Scrub-Jays are mostly blue from head to tail, with white on the undersides and a long tail. This species occurs from the Edwards Plateau to the Pacific Ocean, mostly in dry oak and juniper woodlands. Their call is variable but fairly loud, often a series of harsh *shreep*'s. They are known for their ability to remember where they hide food items such as acorns.

Common Raven

Corvus corax   Order Passeriformes, Family Corvidae

Ravens are large black birds that are often seen soaring over the Hill Country. Their tail is wedge-shaped. The call is a loud, harsh croak. Ravens are predators and scavengers. Ravens and jays are thought to be among the most intelligent animals.

(Photograph by Dustin Wyatt)

Barn Swallow

Swallows and Purple Martin
Order Passeriformes, Family Hirundinidae

Swallows are small streamlined songbirds with pointed wings that give them great maneuverability in the air, where they catch insect prey. They are often seen lined up on telephone wires. Purple Martins (*Progne subis*) are large swallows that are dark purple with a forked tail. Barn Swallows (*Hirundo rustica*) are blue on the back but white underneath and chestnut on the throat. They often build cup-shaped nests on the sides on buildings. Cliff Swallows (*Petrochelidon pyrrho-nota*) are also dark blue above but have a light-colored patch above a square tail. They build gourd-shaped nests out of mud on the walls under bridges or under cliff ledges. Cave Swallows (*P. fulva*) are similar in appearance to Cliff Swallows but have a paler throat. They build cup-shaped nests in the openings of caves, under bridges, and in culverts.

Carolina Chickadee

Poecile carolinensis   Order Passeriformes, Family Paridae

These small birds have a black cap and throat, white cheeks, and gray back. Carolina Chickadees are a common bird of woodlands of the southeastern United States, with the Hill Country forming the southwestern edge of their geographic range. They often come to feeders for seeds and are resident in the Hill Country in all seasons. Their namesake call is a fast *chick-a-dee-dee-dee*, and they also whistle *see-bee-see-bay*.

Black-crested Titmouse

Baeolophus atricristatus   Order Passeriformes, Family Paridae

These small gray birds with a black crest were formerly considered a subspecies of the Tufted Titmouse. Black-crested Titmice occur only in Texas and Mexico. They frequently form flocks with other small birds such as Carolina Chickadees. Their call is a whistled *peter-peter-peter-peter.*

Canyon Wren

Catherpes mexicanus   Order Passeriformes, Family Troglodytidae

The canyons of the Hill Country are at the eastern edge of the range of the Canyon Wren, which occurs throughout the southwestern United States and Mexico. These birds are small and brown like other wrens, but the distinctive song echoing through the canyons can alert you to their presence. The song is often described as a descending cascade of clear notes, becoming slower as the pitch descends.

Carolina Wren
Thryothorus ludovicianus   Order Passeriformes, Family Troglodytidae

Carolina Wrens are common birds of the southeastern United States, reaching their western limit in the Hill Country. They are brown with a white stripe above the eye, long tail usually held upward, and long down-curved bill. The call is a rapidly repeated *teakettle* and various other sounds. Carolina Wrens prefer wooded habitat with dense undergrowth, where they feed on insects and spiders.

(Photograph by H.M. and W.C. Meddaugh)

Bewick's Wren
Thryomanes bewickii   Order Passeriformes, Family Troglodytidae

Bewick's Wren is very similar to the Carolina Wren but has white corners on a longer tail, paler underparts, and a thinner body. The call is quite different as well, with a sequence of clear notes and trills. Bewick's Wren is generally found in more arid areas than the Carolina Wren, but both are abundant throughout the Hill Country.

(Photograph by H.M. and W.C. Meddaugh)

Ruby-crowned Kinglet
Regulus calendula   Order Passeriformes, Family Regulidae

Ruby-crowned Kinglets are common migrants and winter residents in the Hill Country but fly north for summer breeding. This tiny bird is olive-gray with a white wing bar and a black bar just behind it. The ruby crown of the males is usually covered by other feathers. They are usually seen in wooded areas, actively moving through the forest searching for insects among the branches.

Eastern Bluebird

Sialia sialis   Order Passeriformes, Family Turdidae

Eastern Bluebirds have bright blue backs, wings, and tails. The chest and throat are rust-colored, and the belly is white. They are often seen perched on wires, fence posts, or branches in the open. The Edwards Plateau is at the western limit of the range of this species.

American Robin

Turdus migratorius   Order Passeriformes, Family Turdidae

While the American Robin is a well-known summer bird throughout much of North America, in the Hill Country this species is present only in the fall, winter, and spring. These birds often form large flocks in the winter. American Robins have a red chest, gray back, and white throat streaked with black. They feed on earthworms, other small invertebrates, and berries.

(Photograph by Lindsey Wyatt)

Northern Mockingbird
Mimus polyglottos Order Passeriformes, Family Mimidae

Northern Mockingbirds are abundant throughout Texas all year and have been designated the state bird. They are found in a variety of open habitats, including suburban areas. Mockingbirds are gray with white patches on the wings and white on the sides of the tail. They are highly vocal and, as their name implies, are able to mimic a wide variety of birds. Usually a sound is repeated two to six times, and then they switch to a different sound that is similarly repeated.

(Photograph by Chris Harrison)

European Starling

Sturnus vulgaris   Order Passeriformes, Family Sturnidae

European Starlings were introduced to the United States in 1890 from Europe and have spread over the entire country. However, they are primarily found in urban areas. These birds are black with purple or green iridescence in the summer and white spots during the winter. They have a short tail, unlike the other blackbirds with which they may be confused. They nest in cavities and thus may reduce populations of native cavity-nesting birds such as woodpeckers, bluebirds, chickadees, and wrens.

(Photograph by H.M. and W.C. Meddaugh)

Yellow-rumped Warbler

Setophaga coronata   Order Passeriformes, Family Parulidae

More than 40 species of warblers have been recorded in the Hill
Country during spring and fall migration as they move between
southern (usually tropical) wintering grounds and northern breeding
grounds. The Yellow-rumped Warbler is one of the most common in
North America, and many spend the winter in Texas and the southern states. They are gray to pale brown on the back, with yellow on
the rump and on the sides of the chest. Birds in the Hill Country are
mostly the eastern form of the species, formerly considered a separate
species called the Myrtle Warbler.

(Photograph by H.M. and W.C. Meddaugh)

Golden-cheeked Warbler

Setophaga chrysoparia   Order Passeriformes, Family Parulidae

The Golden-cheeked Warbler is endemic to the Edwards Plateau, where it breeds in the summer. It is the only bird species that breeds only in Texas, and it is listed as an endangered species. These birds spend the winter in Mexico and Central America, then arrive in the Hill Country in March and stay until August. They inhabit forests of oak and juniper, where they use strips of bark from large junipers in building nests. Golden-cheeked Warblers have bright yellow cheeks with a dark line through the eye. The song is several notes with a buzzing quality.

Canyon Towhee

Melozone fusca   Order Passeriformes, Family Emberizidae

The Canyon Towhee is a large gray-brown sparrow with a long tail and an orange patch under the base of the tail. These birds are usually seen on the ground in rocky, semiarid country. They are year-round residents, and the Hill Country is at the eastern edge of their geographic range. Three other species of towhees also occur in the Hill Country but do not breed here (Spotted Towhee, Eastern Towhee, and Green-tailed Towhee).

Sparrows
Order Passeriformes, Family Emberizidae

A variety of sparrows are common permanent residents, winter residents, or migrants in the Hill Country. Sparrows are small songbirds with a short, stout beak. They feed on seeds and insects. Most are usually brown with streaks of white and black. In the Hill Country, common species include the Chipping Sparrow, Rufous-crowned Sparrow, Field Sparrow, Lark Sparrow, and White-crowned Sparrow.

Northern Cardinal

Cardinalis cardinalis   Order Passeriformes, Family Cardinalidae

The male Northern Cardinal is bright red with a crest and large orange bill. Females are duller but also have the crest and large orange bill. The bills of juvenile birds are brownish-black. These are common birds and are easily seen because of their large size, bright color, and loud, clear song (*purty-purty-purty* or *cheer-cheer-cheer*). They are resident in the Hill Country throughout the year.

(Photograph by H.M. and W.C. Meddaugh)

Painted Bunting
Passerina ciris   Order Passeriformes, Family Cardinalidae

The male Painted Bunting is unmistakable with its blue head, red chest, and green wings. The female is light yellowish-green. Painted Buntings are summer residents in most of Texas, favoring areas with dense brush. They feed on insects and seeds in low branches and on the ground. Their song is a short series of high-pitched warbles.

(Photograph by H.M. and W.C. Meddaugh)

Red-winged Blackbird

Agelaius phoeniceus   Order Passeriformes, Family Icteridae

Male Red-winged Blackbirds are all black with a patch of red and yellow on the shoulder. Females are brown with streaks of white and black. Red-winged Blackbirds are often seen in large flocks during the winter, especially in open fields, and often with other species of blackbirds. During the summer nesting season, they are usually seen in marshy areas with tall grass. The song is a loud *conk-a-reee* with the last note trilled. This species is thought to be the most abundant bird in North America.

(Photograph by H.M. and W.C. Meddaugh)

Eastern Meadowlark

Meadowlarks
Order Passeriformes, Family Icteridae

Eastern Meadowlarks (*Sturnella magna*) are most easily recognized by the V-shaped black band on the chest, with yellow on the throat and belly. The back is brown with white and black spots. Eastern Meadowlarks occur in grasslands and pastures and are often seen perched on fence posts or power lines. The Western Meadowlark (*S. neglecta*) also occurs in the Hill Country and is difficult to distinguish from the eastern species. The call of the western species is more bubbly compared to the clear whistles of the eastern species. Photographs of the bird and recording of the song may be necessary for definite identification.

Great-tailed Grackle

Quiscalus mexicanus   Order Passeriformes, Family Icteridae

The Great-tailed Grackle and the Common Grackle (*Q. quiscula*) are two of the most abundant birds in the urban areas of the Hill Country. During the winter, huge flocks often take up residence in trees near parking lots or houses. Both birds are more than 12 inches long with glossy black feathers. The main difference is the longer tail of the Great-tailed Grackle. The calls of both species are various but usually include gurgling, squeaking, and creaking sounds. Courtship behaviors are easily observed in the spring on lawns, when males spread their wings and point their bills to the sky.

Brown-headed Cowbird

Molothrus ater   Order Passeriformes, Family Icteridae

Male Brown-headed Cowbirds are black with a brown head, whereas females are all brown. They feed on insects on the ground, especially near cattle. They flock with other blackbirds and grackles in winter. Cowbirds are nest parasites, laying their eggs in the nests of other songbirds, which then raise the chick as their own. Because the cowbird egg hatches sooner and the nestling is larger, this often reduces the survival rate of the host nestlings. Increasing cowbird populations are thought to be a major factor in the decline of many songbirds. They have been recorded in the nests of more than 200 species of birds.

(Photograph by Dustin Wyatt)

House Finch

Haemorhous mexicanus   Order Passeriformes, Family Fringillidae

Male House Finches are red in front and on top of the head, with a light brown stripe through the eye and streaked back and belly. Females are grayish-brown with streaks of white. House Finches feed on seeds and are frequently seen at bird feeders. Their habitat is woodlands and urban areas. They are resident in the Hill Country throughout the year. Originally native to the western part of the United States, they now occur throughout the country.

House Finch

(Photograph by H.M. and W.C. Meddaugh)

Goldfinches
Order Passeriformes, Family Fringillidae

The Lesser Goldfinch (*Spinus psaltria*) is a common breeding bird in the Hill Country. Males are yellow in front and black on the back and head, with white patches on the wings. Females are pale yellow with grayish wings. The American Goldfinch (*S. tristis*) is seen in the Hill Country in the winter, when its plumage is pale yellow, with black wings and white bars across the wing. Both species feed on seeds in open areas.

(Photograph by H.M. and W.C. Meddaugh)

House Sparrow

Passer domesticus   Order Passeriformes, Family Passeridae

House Sparrows were introduced into the United States in the 1800s from Eurasia. They have become among the most abundant birds here, especially in urban and agricultural areas. They are members of the Old World sparrow family, which are not closely related to our native New World sparrows. Males have a black bib, rusty back, and white cheeks. Females are pale brown and gray with dark stripes on the back. The song is a repeated *cheep*, and they also make short chattering calls. They often nest in human-made structures such as behind signs attached to buildings.

CHAPTER SIX Mammals

T HE STATE OF TEXAS HAS 143 SPECIES of native terrestrial mammals, and about half of those occur in the Hill Country. In addition, several species of mammals have been introduced and become
part of the wild fauna. Mammals in the Hill Country range in size
from tiny mice and bats that weigh less than an ounce to feral hogs
and cougars that weigh more than 100 pounds. Most mammals are
nocturnal, so your best chances of seeing them are early in the morning or late in the evening. Mammals are classified into major groups
called orders, which include well-known groups such as bats, rodents,
and carnivores. The term "carnivore" is somewhat confusing because
it refers to a specific order of mammals, but it can also mean any animal that feeds on other animals (meat eaters). Many mammals are
quite wary and unlikely to be seen. However, their tracks in muddy
areas or their scat along trails can signal their presence. The mammals
in this chapter are grouped by order and arranged alphabetically by
genus and species within the order.

Hunting of mammals, especially deer and hogs, is a major economic
activity in the Hill Country. Many ranches make far better profits by
managing white-tailed deer and leasing the land to hunters than they
can make by raising cattle. As you drive along country roads, you will
notice many areas with high fences that contain managed deer herds
or exotic species such as axis deer. Other common exotics include
blackbuck antelope, addax antelope, fallow deer, sika deer, and aou-
dad sheep.

Virginia Opossum (Possum)
Didelphis virginiana   Order Didelphimorphia

Opossums can be recognized by their gray-and-white coloration and long hairless tail. This species is the only native marsupial in the United States. The tail is prehensile, which aids in climbing trees. Opossums are primarily nocturnal but also often seen in the morning or evening. They are most often seen dead on the road. They are scavengers and predators and can be found in all types of habitats in the Hill Country.

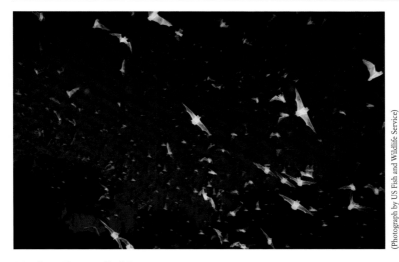

(Photograph by US Fish and Wildlife Service)

Mexican Free-tailed Bat
Tadarida brasiliensis subsp. *mexicana*   Order Chiroptera

This bat, which is a subspecies of the Brazilian free-tailed bat, is the most common and easily observed bat because of the large breeding colonies that occur in Central Texas. There are about a dozen other species of bats in the Hill Country region, but none occur in large colonies like the free-tailed bat. The bat got its name from the long tail that extends far beyond the membrane between the hind legs. Bracken Cave, located in the Hill Country on the northeastern side of San Antonio, is thought to have the largest concentration of mammals in the world, with a colony of up to 40 million Mexican free-tailed bats. These bats are present in caves in the Texas Hill Country from March to October and migrate to the tropics for the winter. In the Hill Country, the following locations provide opportunities for observations of evening bat flights: Bracken Cave (owned by Bat Conservation International), Devil's Sinkhole State Park, Old Tunnel State Park, Kickapoo Cavern State Park, and the Congress Avenue bridge in Austin. Studies have shown that Mexican free-tailed bats consume huge quantities of insects, including many moths whose caterpillars are agricultural pests.

Llano Pocket Gopher
Geomys texensis   Order Rodentia

The Llano pocket gopher is endemic to the Llano Uplift region of the Hill Country and to a small area south of the Hill Country (in Medina, Uvalde, and Zavala Counties). It is closely related to the more widespread plains pocket gopher (*G. bursarius*), which occurs farther north in Texas and the Plains states. These pocket gophers occur only in areas of loamy or gravelly, sandy soil, where they create burrows. The gophers remain in the burrows, pulling plants into the burrows by the roots. Obvious mounds are formed where the gophers push out excavated soil. The pocket in the gopher's name refers to pouches on their cheeks. This species is also known as the Central Texas pocket gopher.

(Photograph by US Fish and Wildlife Service)

Rio Grande Ground Squirrel
Ictidomys parvidens   Order Rodentia

Rio Grande ground squirrels are burrowing squirrels with rows of white spots along the back. They create tunnels within the soil, and the entrances are not mounded. They occur in grassy areas, especially overgrazed pastures and mowed areas. Rio Grande ground squirrels feed on vegetation but will also eat insects and even dead mammals. They hibernate during the cold winter months.

(Photograph by Dustin Wyatt)

Nutria
Myocaster coypus   Order Rodentia

Nutria are introduced aquatic rodents, native to South America. They are smaller than beavers and have a round, ratlike tail, whereas beavers have flattened tails. Nutria are active at night but can also be seen during the day near water. They feed on aquatic vegetation. Beavers (*Castor canadensis*) also occur in the Hill Country but are not common.

Rock Squirrel
Otospermophilus variegatus   Order Rodentia

Rock squirrels are large ground squirrels with bushy tails. They often have black on the head and shoulders. Rock squirrels are usually seen on the ground or in shrubs, while fox squirrels often climb high in trees. Both species are active during the daytime. Rock squirrels on the Edwards Plateau are at the eastern edge of their geographic range, which extends from Colorado and Utah to central Mexico. They prefer rocky terrain and can also be seen on the cliff walls at the San Antonio Zoo and Fiesta Texas in San Antonio.

(Photograph by Chris Harrison)

White-ankled Mouse

Northern Pygmy Mouse
Baiomys taylori

Texas Mouse
Peromyscus attwateri

White-footed Mouse
Peromyscus leucopus

White-ankled Mouse
Peromyscus pectoralis   Order Rodentia

Mice are common mammals throughout the Hill Country but are seldom observed because of their small size and nocturnal habits. Identification to species usually requires close examination of specimens. Trapping studies suggest that the white-ankled mouse is the most common mouse in the Hill Country, or at least the most easily trapped. Several other species of mice also occur in the region, but those listed here are among the most common.

Eastern Fox Squirrel
Sciurus niger   Order Rodentia

The eastern fox squirrel is one of the most easily observed mammals in the Hill Country. This is the common squirrel of woodlands, forests, and cities throughout Central Texas. Eastern fox squirrels are usually reddish-gray in color and have a large fluffy tail. They make large nests of leaves high in trees. They feed on a wide variety of nuts and seeds.

(Photograph by H.M. and W.C. Meddaugh)

Hispid Cotton Rat

Sigmodon hispidus   Order Rodentia

Cotton rats are larger and stockier than mice and have smaller ears. They are most common in grassy areas. Due to their high reproductive rate, the populations of cotton rats can increase rapidly during wet years when much food is available. They feed primarily on grasses and seeds. They are usually nocturnal and are active throughout the year. The nonnative Norway rat (*Rattus norvegicus*) and roof rat (*R. rattus*) are also present in the Hill Country region but usually only in or near grain storage buildings, barns, and garbage dumps.

Armadillos

Nine-banded Armadillo
Dasypus novemcinctus   Order Xenarthra

Armadillos are easily recognized by their bony armor and the noisy way they make their way through leaves and brush. They are commonly seen in campgrounds and along roads throughout the Hill Country and often are a significant part of the roadkill fauna. They feed mostly on insects and are primarily nocturnal. Armadillos can be hosts to the bacterium that causes Hansen's disease, also known as leprosy. Researchers recently found through gene sequencing that some leprosy cases in Texas and Louisiana were probably acquired from armadillos. These most likely were cases where people handled or ate the armadillos. Nine-banded armadillos have litters of four young, which are quadruplets that form from the splitting of a single fertilized egg.

(Photograph by H.M. and W.C. Meddaugh)

Black-tailed Jackrabbit
Lepus californicus   Order Lagomorpha

The black-tailed jackrabbit is the only hare present in Texas. Hares do not nest in burrows as rabbits do. Jackrabbits can be separated from cottontails by their much larger ears. Black-tailed jackrabbits have black on the upper side of the tail and on the tips of their ears. They prefer dry areas with low vegetation such as grasses and cacti. Jackrabbits are well adapted to life in the desert and can survive without drinking, requiring only the moisture in the food they eat. Their large hind legs enable them to run up to 40 miles per hour to escape predators.

Eastern Cottontail

Sylvilagus floridanus   Order Lagomorpha

The abundant and often-seen cottontail rabbit of the eastern United States is also the most common rabbit in the Hill Country. Eastern cottontails are active at night and during the evening and morning, when they can often be seen feeding on vegetation in meadows and lawns. Desert cottontails (*S. audubonii*) also occur in the western part of the Hill Country and are similar to eastern cottontails but with larger ears. Swamp rabbits (*S. aquaticus*) are found only in wet areas.

(Photograph by H.M. and W.C. Meddaugh)

Ringtail
Bassariscus astutus   Order Carnivora

Ringtails are members of the raccoon family (Procyonidae) and are about the size of a domestic cat. They are more slender than raccoons but have a longer tail, which has black-and-white bands like those of raccoons. Ringtails prefer rocky habitat and are active only at night. They are excellent climbers. They are omnivorous, feeding on a variety of smaller animals such as mammals, birds, insects, and reptiles, as well as fruit.

Coyote
Canis latrans   Order Carnivora

Coyotes are more often heard than seen. Their choruses of yips can often be heard after dark in the Hill Country. Although some people find their nighttime noises to be somewhat frightening, coyotes pose no danger to humans. They feed on small rodents, rabbits, and a variety of other food, including fruits. Coyotes and wolves both occurred in the Hill Country until the early 1900s, when a campaign of predator control began to remove them. Both species were extirpated from the region by the 1950s. However, as attitudes toward predators relaxed, coyotes spread to the Hill Country from nearby areas, and they have now completely recolonized the area. Wolves no longer occur in the wild in Texas.

Striped Skunk
Mephitis mephitis   Order Carnivora

There are four species of skunks in Central Texas, but the striped skunk is the most common, especially around human habitations. The striped skunk is black with two stripes on the back that join together at the neck. Skunks are known for their defensive spray that contains a truly awful-smelling sulfur compound. Skunks can be safely observed from a distance, but if the skunk starts stomping or lifts its tail, you should start running. Eastern and western spotted skunks (*Spilogale putorius* and *S. gracilis*) and hog-nosed skunks (*Conepatus leuconotus*) also occur in the Hill Country but are less common. The hog-nosed skunk has a single wide, white stripe down the back.

Northern Raccoon
Procyon lotor   Order Carnivora

Raccoons are easily recognized by their black mask and black rings on the tail. They are the most abundant members of the order Carnivora in Texas. Raccoons are nocturnal and are known for their ability to get into trash cans when foraging. They eat a wide variety of foods, including fruits, seeds, and small animals. They are good climbers.

Common Gray Fox

Urocyon cinereoargenteus   Order Carnivora

Gray foxes are medium-sized members of the dog family. They have a black tip at the end of the tail. Their back is gray with reddish coloration on the sides. They are nocturnal, are seldom seen, and can climb trees. Gray foxes are omnivorous but feed mostly on small mammals. Red foxes (*Vulpes vulpes*) also occur in the Hill Country and can be recognized by their orange-red fur and white-tipped tail.

(Photograph by Carolyn Whiteside)

Axis Deer, Chital
Axis axis (Cervus axis)   Order Artiodactyla

A variety of exotic game species have been introduced to private ranches in the Hill Country and are often observed from the country roads of the region. The most common of these species is the axis deer, also known as the chital. This deer is native to India and adjacent countries. Axis deer have white spots, which enables them to be easily distinguished from white-tailed deer, which lack spots as adults. Axis deer are direct competitors with white-tailed deer.

White-tailed Deer
Odocoileus virginianus   Order Artiodactyla

White-tailed deer are the most easily observed large mammals in the
Texas Hill Country. An evening or morning drive through the coun-
tryside or suburban areas will likely result in several deer sightings. In
many state parks, deer are often seen wandering among the camp-
sites. Research has documented that deer on the Edwards Plateau are
smaller than in other parts of Texas and the United States. Bones of
prehistoric deer from the area show that modern-day deer in this area
are also smaller than prehistoric deer. The decline in large predators
of deer has likely led to an increase in deer population density, leading to food limitation and therefore smaller adult deer. This hypoth-
esis is supported by evidence that in areas of the Edwards Plateau
where deer densities are managed and kept low, the deer are larger.
White-tailed deer are browsers but also graze on herbaceous plants.
Because of the high deer populations, plants that are favorite foods of
deer may be declining.

Javelina, Collared Peccary

Pecari tajacu Order Artiodactyla

Javelinas occur in the western portions of the Hill Country. They prefer areas that contain dense brush for cover. They feed on plants, primarily on prickly pear. Javelinas are members of the peccary family (Tayassuidae), which occurs in North and South America, while feral hogs are members of the Old World pig family (Suidae). Javelinas have long, sharp tusks (canine teeth), so do not approach them.

Feral Pig, Feral Hog, Wild Boar

Sus scrofa   Order Artiodactyla

Feral pigs and fire ants vie for top spot among the most-hated introduced species in Texas. The main effect of feral hogs on people is the damage they cause in their rooting and wallowing. Crop and pasture damage is extensive in the state. In natural areas, grassy vegetation is often torn up by hogs. Feral pigs have large tusks that curve out and upward from the lower jaw. Their abundance has increased dramatically in the Hill Country in the last decade, and land managers now consider them a major problem.

CHAPTER SEVEN **Reptiles**

T EXAS IS HOME TO an amazing diversity of reptiles, many of which are common in the Hill Country. Most of the species found here are widespread in the southeastern United States, the desert Southwest, or the Great Plains. The map turtles are the only endemic reptiles in Central Texas.

Snakes are certainly among the most feared animals in Texas, but fatalities are very rare. There are five species of venomous snakes in the Hill Country, but most snakes here are nonvenomous. Even the venomous species are seldom a threat to humans, but carelessness can result in a dangerous situation. Most snakebites occur when people try to handle, harass, or kill snakes. Snakes feed on small animals and thus are not likely to attack large mammals that are just walking nearby. Watch where you are walking and do not stick your hands or feet into crevices, under rocks, or under logs. Snakes are important predators within the Hill Country ecosystem, and needless widespread destruction of them may result in increased rodent populations. Reptiles in the same taxonomic group are arranged here alphabetically by family and then by genus and species within a family.

Turtles

Snapping Turtle
Chelydra serpentina   Family Chelydridae   Snapping Turtles

Snapping turtles are common over most of the eastern two-thirds of the United States and from Mexico to Ecuador. They are named for their aggressive behavior when encountered on land. The carapace has keels (ridgelike projections) along the middle of the back, and the tail is saw-toothed along the top. Snapping turtles feed on fish, frogs, vegetation, and a wide variety of other food.

Cagle's Map Turtle

Cagle's Map Turtle
Graptemys caglei

Texas Map Turtle
Graptemys versa   Family Emydidae   Box and Water Turtles

Cagle's map turtle is endemic to the Guadalupe, San Antonio, and San Marcos Rivers. The Texas map turtle is endemic to the Colorado River basin (including the Llano River) on the Edwards Plateau. Map turtles feed primarily on insects. These are small turtles, usually less than 8 inches in length. They have keels on the back of their carapace and yellow lines on their heads and legs.

Red-eared Slider
Trachemys scripta subsp. *elegans*
Family Emydidae   Box and Water Turtles

Red-eared sliders are the most common turtle in the Hill Country. The red stripe behind the eye is diagnostic, although older turtles may become so darkly pigmented that the stripe is obscured. The head and legs have yellow stripes, and the carapace is not keeled. These turtles often bask in the sun by climbing out of streams or lakes onto logs or rocks. When approached, they will quickly dive underwater to escape. Red-eared sliders are omnivores, feeding primarily on insects and crustaceans when young, and then gradually shifting to feeding primarily on aquatic plants as they get older. This species is very popular in the pet trade. The Texas river cooter (*Pseudemys texana*) is a related species that also occurs in the Hill Country. It lacks a red stripe on its head, and the lower jaw is flat on the bottom.

Eastern Musk Turtle

Yellow Mud Turtle
Kinosternon flavescens

Eastern Musk Turtle
Sternotherus odoratus   Family Kinosternidae   Musk and Mud Turtles

Musk and mud turtles are small turtles that are often seen crawling along the bottom of a stream or pond. The eastern musk turtle, also known as the stinkpot, has two light lines from the snout to the back of the head. The yellow mud turtle has no lines on the face and a yellow throat. The Mississippi mud turtle also occurs in the Hill Country but appears to be rare. The mud turtles have two hinges on the plastron (bottom shell), whereas musk turtles only have one. Eastern musk turtles are primarily nocturnal and often have algae growing on their carapace. Both species are omnivorous.

Spiny Softshell
Apalone spinifera   Family Trionychidae   Softshell Turtles

Softshell turtles are unlikely to be confused with any other turtle. Their skin is leathery, and they have a pointed snout. Softshells usually occur in rivers and are frequently seen floating at the surface with their nose projecting from the water. These turtles feed on a variety of small animals. They can grow to be 17 inches in carapace length.

Texas Alligator Lizard
Gerrhonotus infernalis   Family Anguidae   Alligator and Glass Lizards

The Texas alligator lizard has large platelike scales, light brown coloration, and jagged white lines across the back. Adults are typically 10–15 inches long. This species occurs in Texas primarily on the Edwards Plateau. Its range also extends south into Mexico. Alligator lizards are usually more slow moving than most other lizards. They feed on insects and small vertebrates.

Eastern Collared Lizard
Crotaphytus collaris
Family Crotaphidae   Collared and Leopard Lizards

Collared lizards have large heads and two black stripes across the back of the neck. The body is green with light spots. Adults can grow to 14 inches long. Their habitat is usually rocky areas, where they may run on their hind legs to escape capture. They feed on insects and small lizards.

Mediterranean Gecko
Hemidactylus turcicus   Family Geckkonidae   Geckos

Mediterranean geckos are abundant on buildings throughout southern Texas. They are nocturnal and congregate around lights and win-
dows that attract insects. Their bodies are pale pink with small dark
spots, large toe pads, and large eyes. They have the ability to make
their tail drop off (tail autotomy) and will do so readily when caught.
The tail will twitch back and forth for about a minute as the gecko
makes its escape. A new tail regrows in a few weeks. This species is
native to the Mediterranean region but is now widespread in cities
from Florida to California. It also can be found on buildings in state
parks in the Hill Country.

Greater Earless Lizard
Cophosaurus texanus
Family Phrynosomatidae   Earless, Spiny, and Horned Lizards

These gray or reddish lizards are about 3–7 inches long. Their coloration usually matches the colors of rocks in their habitat. They have black bands under their tail and often hold their tail curled up. Males have two black bars near the groin. They are most commonly seen on rocky outcrops such as Enchanted Rock and rocky streambeds like those at Pedernales Falls.

Texas Horned Lizard
Phrynosoma cornutum
Family Phrynosomatidae   Earless, Spiny, and Horned Lizards

Also known as horned toads, these lizards are unique with their flattened body and large spines around the back of the head. Their primary food is harvester ants. Populations have declined dramatically in the eastern third of Texas, perhaps due to the spread of fire ants, which prey on harvester ant queens. The Texas horned lizard is listed as threatened on the state endangered species list, and collecting it is illegal.

Prairie Lizard
Sceloporus consobrinus
Family Phrynosomatidae   Earless, Spiny, and Horned Lizards

The prairie lizard is small, usually gray or brown, and typically found in trees or on fences. It usually has light stripes along the body. Males have blue patches on each side of the belly. Their escape behavior is usually to run for a tree and ascend partway up, then dart to the other side of the tree if approached. The prairie lizard is the south-central species in a group of several related species (formerly grouped as *S. undulatus*) that occur widely across the southern United States from Arizona and Utah to the East Coast, from Florida to New York.

Texas Spiny Lizard
Sceloporus olivaceus
Family Phrynosomatidae   Earless, Spiny, and Horned Lizards

Texas spiny lizards are common on tree trunks in the Hill Country. They are often detected when their claws scrape the bark as they run to the other side of the tree or up to safety. They are also sometimes seen on the exterior walls of buildings. They are large lizards with spiny scales and a pale stripe along the side of the body. The crevice spiny lizard (*S. poinsettii*) also occurs in the Hill Country. It is similar to the Texas spiny lizard but has a dark collar above the shoulders and has dark and light bands around the tail. Crevice spiny lizards are usually seen on the ground in rocky areas.

Ornate Tree Lizard
Urosaurus ornatus
Family Phrynosomatidae   Earless, Spiny, and Horned Lizards

Ornate tree lizards are small lizards found both in trees and on rocks.
They are usually gray or tan, with paired dark blotches from shoulder to tail. They lack the spiny scales present in Texas spiny lizards
and fence lizards. Ornate tree lizards have a fold of skin at the base of
the throat. Central Texas is the easternmost part of their range, which
extends west to Utah and Mexico.

Green Anole
Anolis carolinensis   Family Polychrotidae   Anoles

These slender, green lizards are common residents of the Hill Country. Males have large pink throat fans that they use to court females and warn away other males. They are sometimes called chameleons because of their ability to change their skin color from green to brown. Green anoles occur from Central Texas to North Carolina. They are often seen on fences and shrubs.

Ground Skink
Scincella lateralis   Family Scincidae   Skinks

Also known as the little brown skink, this small lizard with short legs is often detected by the rustling it makes in dead leaves. It has brown or coppery smooth scales. Its body moves laterally like a snake's when it runs. This species is common throughout the southeastern United States, reaching its western boundary on the Edwards Plateau. The four-lined skink (*Plestiodon tetragrammus*) also occurs in the Hill Country. It has four light stripes on the head that end at the shoulder in the subspecies found here (the short-lined skink).

Texas Spotted Whiptail

Texas Spotted Whiptail
Aspidoscelis gularis

Six-lined Racerunner
Aspidoscelis sexlineata   Family Teiidae   Whiptails

Whiptails are fast-running lizards with long tails. Both species in our area have dark and light stripes running parallel down the tail and body. The Texas spotted whiptail also has light spots on the dark stripes on the side of the body. These lizards usually inhabit open areas such as grasslands or riparian areas. They escape rapidly if approached by entering dense vegetation or burrows.

Texas Ratsnake

Ratsnakes

Pantherophis (*Elaphe*) spp.   Family Colubridae   Colubrids

There are three species of ratsnake commonly seen in the Hill Country. The Texas ratsnake (*Pantherophis obsoletus*) is yellow or grayish with dark blotches on the back. This species occurs throughout eastern Texas and other parts of the central United States. The Great Plains ratsnake (*P. emoryi*) has brown blotches on a gray background and a light spearpoint pattern on the top of the head behind the eyes. It occurs throughout the south-central region of North America. Baird's ratsnake (*P. bairdi*) is gray or light brown with four longitudinal darker stripes. This species occurs in the western part of the Hill Country and farther west in Texas. All of the species have a keel (ridge) in the middle of each scale on the back. Ratsnakes can grow to 6 feet in length and thus are often feared. However, they are not venomous or aggressive and are important predators of small rodents. They are good climbers and occasionally seen in trees.

(Photograph by Forrest M. Mims III)

Diamond-backed Watersnake

Watersnakes
Family Colubridae   Colubrids

Watersnakes are large, aquatic snakes that are often confused with cottonmouths. Water snakes are nonvenomous, although they will bite vigorously if handled. The plain-bellied watersnake (*Nerodia erythrogaster*) has dark patches along the back, alternating with dark patches along the sides, with a lighter brown background. The diamond-backed watersnake (*N. rhombifer*) also occurs in the Hill Country but has a dark chain-link pattern along the back, with lighter areas in between.

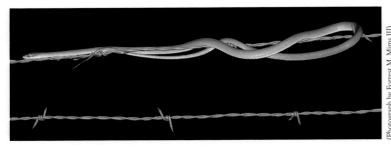

(Photograph by Forrest M. Mims III)

Rough Greensnake
Opheodrys aestivus   Family Colubridae   Colubrids

Because of its color, this snake is unlikely to be confused with any other in the Hill Country. Rough greensnakes are usually about 2–3 feet long with a thin body. They are excellent climbers and spend much of their time in trees and shrubs, where they are well camouflaged. They feed primarily on insects and spiders.

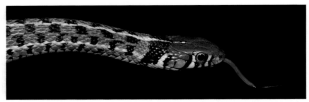

Checkered Gartersnake

Gartersnakes
Thamnophis spp.   Family Colubridae   Colubrids

Gartersnakes are long, thin snakes with a distinct stripe down the middle of the back. In the Hill Country, three species are fairly common. The black-necked gartersnake (*T. cyrtopsis*) has an orange stripe down the middle of the back and light yellow stripes on either side, a large black spot behind the head, and a series of black spots continuing down the neck. The checkered gartersnake (*T. marcianus*) is very similar to the black-necked gartersnake, with a pale yellow stripe on the back, a black spot behind a lighter band behind the head, and a distinct checkerboard pattern of black spots down the back. The western ribbonsnake (*T. proximus*) is all stripes without the spotted pattern of the previous two species. The stripe on the middle of the back is usually red in western ribbonsnakes from the Hill Country.

Rough Earthsnake
Virginia striatula   Family Colubridae   Colubrids

This small snake lacks distinctive markings. The top is usually light brown to gray, and the underside is pale. The head is cone-shaped, and scales are keeled. Earthsnakes feed on earthworms and other small invertebrates. They do not bite if handled. They occur in a variety of habitats from Central Texas to Virginia.

(Photograph by Forrest M. Mims III)

Texas Coralsnake   VENOMOUS
Micrurus tener   Family Elapidae   Coralsnakes and Cobras

The Texas coralsnake is highly venomous and should not be handled. The venom is more toxic than that of the vipers in Texas. Texas coralsnakes have rings of black, yellow, and red encircling their body. The snout is black. Texas coralsnakes feed on snakes and lizards. If left alone, they pose no threat to humans.

Broadbanded Copperhead

Copperhead   VENOMOUS
Agkistrodon contortrix

Cottonmouth   VENOMOUS
Agkistrodon piscivorous   Family Viperidae   Vipers

Members of the viper family are stout snakes recognized by their triangular head, small pits below the eyes, and pupils that are vertical slits. All are venomous and should not be disturbed, handled, or approached. Copperheads are beautiful snakes with distinct dark and light copper-colored bands. They are well camouflaged on a background of dead leaves. Young copperheads have yellow on the tip of their tail. Cottonmouths, sometimes called water moccasins, are closely related vipers that are usually seen in or near water. They have a mottled pattern of light and dark irregular blotches. Older individuals are often completely dark. Their defensive posture is to hold the mouth open wide, showing the white lining.

Western Diamond-backed Rattlesnake   VENOMOUS
Crotalus atrox   Family Viperidae   Vipers

Western diamond-backed rattlesnakes are the largest snakes in Texas, often 3–4 feet long with thick, heavy bodies. They have brown or gray scales, with a diamond pattern along the back, which may not be very distinct. There are two light diagonal stripes on each side of the head and black rings around the tail. The rattle is made of keratin, the same substance as snake scales and human hair and fingernails. Contrary to myth, rattlers do not add an extra rattle each year. Instead, they add a rattle each time they shed their skin. This may occur several times a year, but rattles also frequently break off, so the number of rattles does not indicate anything about the age of the snake. Rattlesnakes are important predators of small mammals, especially rodents.

Most of the venomous snakebites in Texas are due to western diamondbacks. Young rattlesnakes actually have more toxic venom than older snakes. Opossums are remarkably resistant to rattlesnake venom and will eat rattlesnakes, although recent research suggests that the venom of rattlesnakes may be evolving rapidly to protect them against opossums. Backing away is the best strategy when encountering a rattlesnake.

The rock rattlesnake (*C. lepidus*) is a smaller rattlesnake that usually has crossbars across the back end of the body, while the black-tailed rattlesnake (*C. molossus*) can be recognized by its black tail. Both of these species occur from the Hill Country to Arizona and south into Mexico, while the western diamond-backed rattlesnake's range extends from Oklahoma and Arkansas west to California and south to Mexico.

CHAPTER EIGHT Amphibians

DESPITE ITS FAIRLY ARID CLIMATE, the Texas Hill Country is home to a remarkable variety of amphibians. This species diversity is concentrated in and near the springs, streams, and groundwater of the Hill Country. As in the Galapagos Islands, where individual islands often have their own species of finch or tortoise, many of the isolated springs and small streams of the Hill Country also have unique species of salamanders. In addition to these fully aquatic species, the cliff chirping frog is endemic to the Edwards Plateau.

Unlike the scaly skin of reptiles, amphibian skin is moist and permeable to water. Because they will rapidly lose water in dry conditions, amphibians are usually active at night or in rainy weather. They are most easily observed by walking roads or trails after a heavy rain or by walking along streams or ponds.

One of the easiest ways to detect the presence of frogs and toads in a region is to listen for their calls on rainy nights. Male frogs and toads make loud calls to attract females. These calls are unique to each species, and by learning the calls, you can identify which species are present even without seeing them. The calls can also be used to find frogs and toads if you want to observe them. Recordings of frog and toad calls are available on compact discs and websites.

The amphibians are grouped into "salamanders" and "frogs and toads." Each group is arranged alphabetically by family and then alphabetically by genus and species within a family.

Salamanders

Texas Blind Salamander

Stream-Dwelling and Aquifer-Dwelling Salamanders
Eurycea spp.   Family Plethodontidae   Lungless Salamanders

Within the cool, spring-fed streams of the Hill Country and the Edwards Aquifer a large number of unique salamanders have evolved in watersheds that are isolated from one another by the warm conditions between them. It is not yet known how many different species in the genus *Eurycea* inhabit these small streams, but more than 10 species have already been named. Many are found in only a single stream or spring, while others live in the water underground. Most could be considered endangered because of the small area in which they are found, although they may be abundant locally. These salamanders remain aquatic throughout their lives and can be recognized by the feathery gills on the sides of their heads. The Texas blind salamander (*E. rathbuni*) is found only in the aquifer below the city of San Marcos in Hays County.

Western Slimy Salamander

Plethodon albagula   Family Plethodontidae   Lungless Salamanders

Western slimy salamanders are black with white spots. When handled, they usually release an extremely sticky skin secretion from their tails. These salamanders do not have lungs: they obtain all their oxygen through their skin. They are usually found around springs, seeps, or caves. They lay their eggs on land and do not have a larval stage.

Coastal Plain Toad

Gulf Coast Toad, Coastal Plain Toad
Incilius (Bufo) nebulifer   Family Bufonidae   True Toads

The Gulf Coast toad is the most commonly seen toad in the Hill Country. It is a large toad with a distinct white line down its back and white stripes on the sides. It has distinct ridges called cranial crests on its head. The mating call is a short trill, lasting two to six seconds. Coastal Plain toads live in a variety of habitats, including suburban areas, where they often are seen near lights or hopping across roads. Four other species of related toads also occur in the Hill Country: green toad, *Anaxyrus (Bufo) debilis*; red-spotted toad, *A. (Bufo) punctatus*; Texas toad, *A. (Bufo) speciosus*; and Woodhouse's toad, *A. (Bufo) woodhousei*.

Blanchard's Cricket Frog
Acris blanchardi (*A. crepitans blanchardi*)   Family Hylidae   Treefrogs

Blanchard's cricket frogs are very common along the shoreline of
most water bodies in the Hill Country. They are small frogs that are
easily overlooked but also easily detected by their call, which sounds
like two rocks being clicked together repeatedly. They have warty
skin, with a dark triangle behind the eyes and dark bands across the
tops of their thighs.

Gray Treefrog

Cope's Gray Treefrog
Hyla chrysocelis

Gray Treefrog
Hyla versicolor   Family Hylidae   Treefrogs

These two species are indistinguishable by appearance. Gray treefrogs have twice as many chromosomes as Cope's gray treefrogs, and their mating call is a slower trill. The color of their skin is variable and is somewhat warty compared to that of other treefrogs. The hidden surfaces of their hind legs are bright yellow. They have large toe pads and are usually found in trees. These frogs are at the western edge of their distribution, which covers most of the eastern United States.

Green Treefrog
Hyla cinerea   Family Hylidae   Treefrogs

Green treefrogs are bright green with a white stripe down each side. They have large toe pads that enable them to easily climb in trees or straight up the walls of buildings. They are sometimes found near lights on buildings as they wait for insects. Their call is a short, nasal, bell-like sound, repeated every second. Green treefrogs are common in the Hill Country and throughout the southeastern United States.

(Photograph by Gary Nafis)

Strecker's Chorus Frog

Pseudacris streckeri   Family Hylidae   Treefrogs

This small frog (less than 2 inches long) is usually brown or greenish and has a dark stripe from the snout, through the eye, to the shoulder. These frogs are found in a variety of habitats. They mate in the winter. Their call is a short, bell-like note that is rapidly repeated. The spotted chorus frog (*P. clarkii*) also occurs in the Hill Country. It is a bit smaller than Strecker's chorus frog, has distinct green spots on the back, and has a call that is a short, scraping trill, rapidly repeated.

(Photograph by Chris Harrison)

Cliff Chirping Frog
Eleutherodactylus marnockii
Family Leptodactylidae   Neotropical Frogs

The cliff chirping frog is a small frog, less than 1.5 inches in length, of yellow-green coloration with dark spots. Its head is large in comparison to the rest of its body. It is found only in Texas, on the Edwards Plateau and farther west on the Stockton Plateau. Cliff chirping frogs are nocturnal. Their call is a short, high-pitched chirp like that of a cricket or a short, rattling trill. This species lays its eggs in moist areas on land, and they develop directly into frogs within the egg, bypassing the aquatic tadpole stage that occurs in the other frog families of Texas. The parents guard the eggs until they hatch.

Western Narrow-mouthed Toad

Gastrophryne olivacea   Family Microhylidae   Microhylids

Western narrow-mouthed toads have triangular, pointed heads and round bodies with short legs. They are adapted for burrowing into loose soil and feed on ants and termites. Their calls sound like the bleat of a lamb. This species ranges from southern Nebraska to Mexico.

Rio Grande Leopard Frog
Lithobates (Rana) berlandieri   Family Ranidae   True Frogs

Leopard frogs are aquatic so are usually found in or near water but sometimes are found away from water during or after rains. They have tan or gray skin, with dark blotches over most of the body and dark bands across the hind legs. A fold of skin runs down both sides of the back, and near the hind legs these folds are interrupted and inset toward the middle. Their call is a snorelike rattle. Bullfrogs (*L. catesbeianus*) also occur in the Hill Country. They are similar to leopard frogs but can get much larger and lack the distinctly shaped blotches. They do not have skin folds along the sides of the back. Their call is usually described as a low-pitched *jug-o-rum*.

(Photograph by Chris Harrison)

Couch's Spadefoot Toad

Scaphiopus couchii   Family Scaphiopidae   Spadefoot Toads

Spadefoot toads have a black tubercle on their hind feet that helps them dig backward into soil. Couch's spadefoot has greenish or yellowish skin with indistinct darker areas. The call is a harsh bleat, repeated every five seconds. The tadpole stage is very short, usually lasting only about two weeks.

CHAPTER NINE Fishes

THE HILL COUNTRY IS A POPULAR DESTINATION for recreational fishing, with a variety of habitats ranging from fast-flowing streams to deep reservoirs. There is even a seasonal fishery of rainbow trout, which are stocked each winter into the Guadalupe River below Canyon Dam when the water temperature is low enough for them to survive. In addition to several species of introduced sport fish, a wide variety of interesting native fish can be observed from the banks of clear streams or by snorkeling. There are several species of fish that are unique to the Hill Country. The fishes are listed in taxonomic groups, arranged alphabetically by family within each group and then by genus and species within a family.

Spotted Gar

Lepisosteus oculatus   Family Lepisosteidae   Gars

Gars are found throughout Texas in backwaters and stream pools. Spotted gar and longnose gar (*L. osseus*) have spots on the body and tail, but the spotted gar also has spots on its head. Longnose gars have a long, narrow snout that is 10 times longer than its narrowest width. Longnose gar can grow up to 80 pounds, but record sizes for Hill Country lakes are usually 10–30 pounds. There are seven species of gars in North America, but only these two in the Hill Country. They are piscivores (fish eaters).

(Photograph by Sam Stukel)

Gizzard Shad

Gizzard Shad

Dorosoma cepedianum

Threadfin Shad

Dorosoma petenense   Family Clupeidae   Shads

Shads are small to medium-sized fish that are commonly used as bait for fishing. Their habitat in the Hill Country is the open-water areas of reservoirs. They have a compressed body and saw-toothed scaling along the belly. Gizzard shad have a blunt snout with the mouth slightly below the end, while threadfin shad have a pointed snout with the mouth opening at the point.

Common Carp

Cyprinus carpio   Family Cyprinidae   Carps and Minnows

Common carp is an introduced species from Eurasia that is now widespread in lakes and streams in Texas. These fish have a deep, thick body, usually golden or greenish-yellow, with two barbels on each side of the jaws. They favor muddy areas and are known for stirring up the sediments. They often gulp air at the water's surface.

Minnows

Blacktail Shiner

Central Stoneroller
Campostoma anomalum

Red Shiner
Cyprinella lutrensis

Blacktail Shiner
Cyprinella venusta

Texas Shiner
Notropis amabilis

Sand Shiner
Notropis stramineus   Family Cyprinidae   Carps and Minnows

These five species are just a sample from the wide variety of minnows found in Central Texas. Surveys by fish biologists document that minnows are the most abundant fish in the rivers of the Hill Country. While many people call any small fish a minnow, for biologists the term refers to members of the Cyprinidae, which share some common anatomical features, such as a single dorsal fin, lack of teeth on the jaws, and presence of teeth on bones in the throat (called pharyngeal teeth).

Suckers

Smallmouth Buffalo

River Carpsucker
Carpiodes carpio

Smallmouth Buffalo
Ictiobus bubalus

Gray Redhorse
Moxostoma congestum   Family Catostomidae   Suckers

The mouth of suckers is on the bottom of the head, enabling them to suck food off the bottom more easily. River carpsuckers have large eyes and large scales. The dorsal fin is much longer in the front than at the back. Gray redhorses have thick lips with folds or grooves. Smallmouth buffalo have a dorsal fin similar to that of the river carpsucker and distinct grooves on the upper lip.

Channel Catfish

Black Bullhead
Ameiurus melas

Yellow Bullhead
Ameiurus natalis

Blue Catfish
Ictalurus furcatus

Channel Catfish
Ictalurus punctatus   Family Ictaluridae   Bullhead Catfishes

Catfish have eight barbels, fleshy projections that sense taste and touch, around their mouth. Catfish lack the bony scales that most other fish have. The bullheads have straight-ended tail fins, while the catfish have forked tail fins. The body coloration of the two bullhead species is similar, but the black bullhead has black barbels, while those of yellow bullheads are white or yellow. Blue and channel cats can be separated by the anal fin (at the bottom rear of the body), which is straight in blue cats and rounded in channel cats. Channel cats often have dark spots on their sides.

Inland Silverside
Menidia beryllina   Family Atherinidae   Silversides

Inland silversides are native to the coastal rivers of Texas but are common in the reservoirs of the Hill Country due to introductions. These small minnowlike fish have a silvery band along the side, a short snout, large eyes, and a forked tail fin. They can be seen near the surface in shallow water, often in large schools.

Western Mosquitofish
Gambusia affinis   Family Poeciliidae   Livebearers

Although mosquitofish are often mistaken for minnows because they are small and common in lakes and streams, they are actually members of the same family as guppies and mollies. Mosquitofish have a rounded tail fin, a flat-topped head, and a dark blotch under each eye. Their name comes from their feeding on the aquatic larval stage of mosquitoes. They have been widely introduced into water bodies to control mosquito populations. Two other species of gambusia are native to the Hill Country, but both are rare endemics.

White Bass
Morone chrysops   Family Moronidae   Temperate Basses

White bass were introduced to the reservoirs of the Texas Hill Country as a sport fish. They are gray to silver with thin, dark, horizontal stripes on the upper parts of their sides. They have two separate dorsal fins: the front one has spiny rays, and the back one has soft rays. They are closely related to striped bass (*M. saxatilis*), which are stocked in some of the Hill Country reservoirs. Striped bass have a longer body and two tooth patches on the tongue, whereas white bass have only a single tooth patch.

Sunfishes

Bluegill

Redbreast sunfish
Lepomis auritus

Green sunfish
Lepomis cyanellus

Bluegill
Lepomis macrochirus

Longear sunfish
Lepomis megalotis

Redear sunfish
Lepomis microlophus
Family Centrarchidae   Black Basses and Sunfishes

Sunfishes have compressed bodies and two dorsal fins that are joined together, appearing as a single fin. They have three sharp spines in the anal fin. A good field guide to fishes is needed to identify the species, and even then there may be difficulties because sunfish species often interbreed, producing hybrids. Sunfishes are often called perch in Texas, but this name is better restricted to members of the perch family. Sunfishes are common in nearly all types of aquatic habitats in the Hill Country.

Basses

Largemouth Bass

Smallmouth Bass
Micropterus dolomieu

Largemouth Bass
Micropterus salmoides

Guadalupe Bass
Micropterus treculii

White Crappie
Pomoxis annularis   Family Centrarchidae   Black Basses and Sunfishes

These fishes are more closely related to sunfishes than they are to white bass or striped bass. Like the sunfishes, they have two dorsal fins joined together, but their body is more elongate, the mouth is larger, and they can grow to a much larger size. The largemouth bass has an upper jaw that extends back past the eye and a dark horizontal stripe down the middle of the side. The Guadalupe bass is endemic to Central Texas and is the official state fish. It has a smaller mouth and 10–12 dark vertical bars along its side. The smallmouth bass is introduced to this region and has three dark lines on each cheek. Guadalupe bass and smallmouth bass have hybridized, a factor threatening the existence of the Guadalupe bass.

Darters

Orangethroat Darter

Greenthroat Darter
Etheostoma lepidum

Orangethroat Darter
Etheostoma spectabile

Texas Logperch
Percina carbonaria   Family Percidae   Perches and Darters

Darters are small fish that are usually found just above the gravel bottom of riffle habitats in streams. The males of many species are brightly colored during the breeding season. The three species listed here are the most common in the Hill Country. The fountain darter (*E. fonticola*) is endemic to San Marcos and Comal Springs.

Rio Grande Cichlid
Herichthys cyanoguttatus   Family Cichlidae   Cichlids

The Rio Grande cichlid is native to the Rio Grande and Pecos River of Texas but not to the Hill Country. However, it has been widely introduced and is now commonly found here in reservoirs and rivers. This species has many small white spots covering its body and a long, single dorsal fin. Maximum size is 12 inches long.

CHAPTER TEN

Invertebrates

THOUSANDS OF SPECIES of invertebrates inhabit the land and water habitats of the Hill Country. This chapter depicts some of the larger, most obvious, or most interesting species. The most unique species are those that live in the caves and springs of the Hill Country. Many of these species are endangered because they occupy such a small habitat that damage to that habitat could lead to extinction. Several types of aquatic invertebrates are included here because of the importance of water quality in the Hill Country. Mayflies, riffle beetles, and caddisflies are important indicators of healthy streams. Many of these animals are highly sensitive to water pollution or to the effects of pollution on oxygen concentrations in the water. Thus, the presence of these species in a stream can indicate a lack of pollution. The species are listed in taxonomic groups, with each group arranged alphabetically by class, order, or family, and then by genus and species within a family.

Asian Clam
Corbicula fluminea   Family Corbiculidae   Basket Clams

This clam has a thick shell with distinct ridges, grows to about 2 inches in diameter, and is usually brown in color. It is an introduced species that has become abundant throughout rivers in the Hill Country. Clams are filter feeders, consuming small particles such as algae and dead plant matter.

Unionid Mussels
Family Unionidae   Freshwater Mussels

This group of clams contains all of the large native species in Texas. Most are oval in shape, with a range of colors from yellow to brown. Some species grow up to several inches in diameter. About 50 species occur in the state of Texas, but many have declined dramatically due to dams, habitat loss, and water pollution. Several unionid mussels are on the state's list of endangered species.

Yellow Garden Spider

Argiope aurantia   Class Arachnida   Arachnids
Order Araneae   Spiders
Family Araneidae   Orb Weavers

This large yellow-and-black spider has long legs that are yellow toward the base and black toward the ends. The females are about an inch in length and create large webs used to trap grasshoppers and butterflies. The males are only about ¼ inch and are usually observed at the edge of the female's web, generally in the late summer or fall. The females produce a large brown egg sac filled with hundreds of eggs.

Tarantula
Aphonopelma spp.
Order Araneae   Spiders
Family Theraphosidae   Tarantulas

Tarantulas are large brown-and-black spiders with hairy legs. They create burrows or use existing cavities, where they wait for insect prey to walk by. They can bite if disturbed, but the venom is not fatal to humans. They also have stinging hairs on their abdomen. Males and females are 7–10 years old before they become mature, and female tarantulas have been known to live 25 years in captivity. The taxonomy of the genus is poorly understood, and species are difficult to identify.

Harvestmen
Order Opiliones

Harvestmen, or daddy longlegs, are not spiders. Unlike spiders, they lack venom and silk glands. They have a round body, long legs, and only two eyes, in contrast to the eight eyes of spiders. They are predators and scavengers on insects and plant juices. In the Hill Country, they are often seen in large aggregations on rock walls and in caves.

Striped Bark Scorpion
Centruroides vittatus
Order Scorpiones   Scorpions
Family Buthidae   Buthid Scorpions

This scorpion occurs throughout Texas and is common in the Hill Country under rocks and wood. It is often seen on exterior walls of buildings or on the floor inside. These scorpions have two broad stripes on the back. The sting is not fatal but causes pain and swelling like a bee sting.

Millipede

Centipedes
Class Chilopoda

Millipedes
Class Diplopoda

Centipedes have one pair of legs per body segment, and millipedes have two pairs per segment. Centipedes are typically fast-running predators, whereas millipedes are slow-moving herbivores. Centipedes have poison glands and fangs and should not be handled. One of the most awe-inspiring invertebrates in the Hill Country is certainly the giant red-headed centipede (*Scolopendra heros*). These grow up to 8 inches in length and have black body segments, a red head, and yellow legs.

Crustaceans

Fairy Shrimp
Order Anostraca

Fairy shrimp are small crustaceans, less than an inch long, that dwell only in temporary pools. They have numerous small legs and swim upside down. The shallow pools on top of Enchanted Rock often have abundant populations of fairy shrimp. Fairy shrimp live in temporary pools because these lack fish, and the fairy shrimp have no defenses against fish predation.

Crayfish
Order Decapoda

Crayfish are the freshwater relatives of lobsters. They have five pairs of legs, and the front two or three pairs have claws at their ends. Crayfish are omnivores, and they live in most types of aquatic habitats in the Hill Country.

Pillbugs
Order Isopoda

Pillbugs are terrestrial crustaceans that look somewhat like tiny armadillos. Their typical defense is to roll into a tight ball, giving them the common name of roly-poly. They commonly enter buildings and are a nuisance but not harmful. They feed primarily on decaying vegetation.

Cockroaches
Class Hexapoda (Insecta)   Order Blattodea

Cockroaches have flattened oval bodies, usually dark brown. The wings are held flat against the back. Cockroaches occur in natural habitats in Texas, where they are decomposers of organic matter. But they are better known for those species that live in houses, feeding on food scraps.

Riffle Beetles
Order Coleoptera Beetles
Family Elmidae   Riffle Beetles

These tiny black or brown beetles are common in the spring-fed streams of the Hill Country. They usually occur on rocks in riffles, the shallow and fast-moving portions of streams. They are less than ¼ inch in length. The Comal Springs riffle beetle is an endangered species that occurs only in Comal Springs and San Marcos Springs.

Dung Beetles
Order Coleoptera Beetles
Family Scarabaeidae   Scarab Beetles

Dung beetles feed on the feces of large mammals, and horseback riding trails in the Hill Country are good places to find these interesting beetles. The beetles create a ball of fresh dung and then roll it with their hind legs. The dung is buried and eggs are laid on the dung, which is then consumed by the larvae.

Pinacate Beetle, Stink Beetle

Eleodes spp.   Order Coleoptera   Beetles
Family Tenebrionidae   Darkling Beetles

Pinacate beetles are fairly common in the arid western United States, including the Hill Country. They are black, shiny beetles, about an inch in length, that are usually seen walking on the ground. They have a distinctive posture in which they elevate the hind end at an angle of about 45°. They emit a foul odor if handled.

Earwigs
Order Dermaptera

Earwigs can be recognized by the pincers on the back end of the abdomen, which do not sting. They have a narrow brown body, with small folded wings on adults (although they seldom fly). Earwigs are nocturnal and feed on other insects and decaying vegetation.

Leafminer trails

Leafminer Flies
Order Diptera   True Flies
Family Agromyzidae   Leafminer Flies

Leafminer flies are small and inconspicuous, but the trails of their larvae are obvious on many leaves, especially frostweed. The flies lay their eggs on the leaves, and then tiny larvae burrow through the leaves as they eat the plant tissue. This creates a winding white line in the leaf. Notice that the tunnel grows wider as the larva gets larger. After about a week, the larva cuts a hole out of the leaf and drops to the ground, metamorphoses to the pupa stage, and then turns into an adult fly.

Black Flies
Order Diptera   True Flies
Family Simuliidae   Black Flies

The larvae of black flies are abundant in Hill Country streams in very specific conditions. Look on rocks in waterfalls or on dams where the flow of water is a smooth, thin layer. There you will see a large number of small black larvae that are stuck to the rock surface. These black fly larvae are filter feeders, with large, fanlike mouthparts used to capture algae and other particles from the flowing water. Adult black flies are small and seldom bothersome in Texas but are very abundant in early spring in the boreal forest of northern North America.

Mayflies
Order Ephemeroptera

Mayflies are small insects with aquatic larvae and short-lived flying adults. The larvae are narrow with tiny leafy or feathery gills along the abdomen. Mayflies are an important food source for fish, and the flies used by fly fishermen often mimic particular types of mayflies. The larvae can be found by lifting stones from the streambed and looking on the bottom and sides. Mayfly adults are usually seen in spring or summer when flying near water after metamorphosing from the larval stage. About 15 different genera have been found in Hill Country streams.

Cicadas

Order Hemiptera   True Bugs
Family Cicadidae   Cicadas

Cicadas are large green insects that are extremely loud. Summer afternoons in the Hill Country are dominated by the whine of the male cicada's call, which is produced by vibrating membranes on its abdomen. The nymphs of cicadas feed on sap from tree roots and then crawl up on the trunk or branch of a tree before molting to become adults. The dry brown skin of the last stage of the nymph is perhaps easier to find than the adults, which are well camouflaged.

Cochineal Bugs
Order Hemiptera   True Bugs
Family Dactylopiidae   Cochineal Bugs

Cochineal bugs are common on prickly pear cactus, where the females cover their body in a white waxy material. They have piercing mouthparts that enable them to feed on the cactus. Crushing the bug yields a dark red dye, which was widely used by Native Americans and early settlers. The males look like small gnats and are seldom seen.

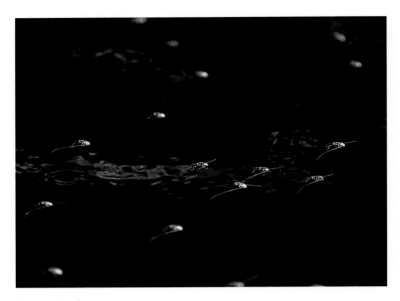

Water Striders
Order Hemiptera   True Bugs
Family Gerridae   Water Striders

Water striders live on the surface of streams and ponds. They have many tiny hairs on their feet that enable them to stand on the surface tension of the water surface. They prey on small insects that fall onto the surface of the water and get stuck in the surface tension. Members of the family Gerridae are usually about ½ inch in body length, while much smaller water striders belong to other related families.

Creeping Water Bugs
Order Hemiptera   True Bugs
Family Naucoridae   Creeping Water Bugs

Creeping water bugs are common aquatic bugs in the streams of the Hill Country. They are most easily seen by lifting rocks in shallow, flowing water. They are typically less than ½ inch in length, brown, with an oval to round body, flattened, and with very wide front legs (in the section analogous to our biceps). They feed on invertebrates by piercing them with their sharp, beaklike mouth. Giant water bugs (family Belostomatidae) also occur in the Hill Country but are much larger (up to 2 inches).

Bumblebee

Bees
Order Hymenoptera   Ants, Bees, and Wasps
Superfamily Apoidea   Bees and Sphecoid Wasps

Bees are familiar components of the insect fauna worldwide. They usually can be recognized by their hairy body and attraction to flowers, and they are important pollinators for a wide variety of plants. Their wings are clear or dark with few veins, the forewing is larger than the hindwing, and the forewings and hind wings are usually attached to each other by tiny hooks. More than 300 species of bees have been recorded in Austin alone. Bumble bees (*Bombus* spp.) are large bees with hairy abdomens colored in yellow and black. Their nests are usually underground in cavities such as abandoned rodent burrows. Honeybees (*Apis mellifera*) are the bees that are kept by bee-keepers in aboveground hives. They are an introduced species from Europe, and some have interbred with an introduced African strain to form the so-called killer bees, which are more aggressive than the European strain. Both bumblebees and honeybees are social insects, living in colonies, but most other bees are solitary. Only female bees can sting.

Galls of mealy oak gall wasp on live oak leaves

Mealy Oak Gall Wasp

Disholcaspis cinerosa

Order Hymenoptera   Ants, Bees, and Wasps

Family Cynipidae   Gall Wasps

Although this wasp is seldom seen, the galls that its larvae create on live oaks are common and easily noticed. The wasp lays its eggs on leaf buds, and the presence of the larvae stimulates the growth of plant tissue around them to form the gall. The galls are about an inch in diameter when fully formed. After the larvae go through a pupal stage in the gall, the adult wasp emerges. The galls have no apparent effect on the health of the trees. A related species called the oak apple wasp produces large galls on red oaks.

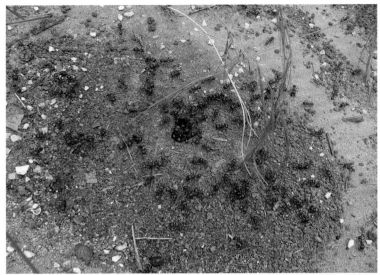
Mound of red harvester ants

Red Harvester Ants
Pogonomyrmex barbatus
Order Hymenoptera   Ants, Bees, and Wasps
Family Formicidae   Ants

Harvester ants are fairly large ants, up to ½ inch long, and are red or brown. Their colony opening is usually a hole in the ground surrounded by a circle of bare ground a few feet in diameter and trails radiating out from the entrance hole. The ants collect seeds and store them in the underground nest. The ants sold for "ant farms" are usually the workers of this species. They can sting but are not aggressive unless bothered. Harvester ants are the primary prey for horned lizards.

Mound of red imported fire ants

Red Imported Fire Ants
Solenopsis invicta
Order Hymenoptera   Ants, Bees, and Wasps
Family Formicidae   Ants

The red imported fire ant is one of the most important nonnative species in Texas. These ants are omnivorous, feeding primarily on other insects but also on larger animals. They are widely known because of their aggressive behavior and distinctive conical mounds. When the mound is disturbed, hundreds of fire ants swarm out and attack. Fire ant workers are about ⅛ inch long and reddish in color. Their sting is painful and can cause allergic reactions or become infected. There are also native species of fire ants in Texas. A species of fly that kills red imported fire ants has been released near Austin as part of research on biological methods to reduce fire ant populations.

Tunnel of desert termites

Termites
Order Isoptera

Termites are social insects known for their ability to eat wood, as a result of symbiotic microbes in their gut. Worker termites are small white insects with soft bodies. In the spring, winged males and females form swarms, usually after rains. Subterranean termites (*Reticulitermes* spp.) nest underground or in cavities in wood and are the most destructive to buildings. Desert termites (*Gnathamiterme* spp.) create mud tunnels on dead grass and other small plants. Breaking open these tunnels reveals the workers feeding on the plant material.

Gulf Fritillary
Agraulis vanillae
Order Lepidoptera   Butterflies
Family Nymphalidae   Brush-footed Butterflies

These bright orange butterflies have black markings on rather long wings (3 inches). The underside is a lighter orange with silver spots. Gulf fritillaries are related to the tropical longwing butterflies (Heliconiinae). Orange and blue caterpillars feed on leaves of passionflower plants.

Hackberry Emperor
Asterocampa celtis
Order Lepidoptera   Butterflies
Family Nymphalidae   Brush-footed Butterflies

This butterfly has orange to brown wings with black spots and two parallel lines along the hind wing. The forewing has white spots. The caterpillars are green with two horns on their head and feed on hackberry plants.

Bordered Patch
Chlosyne lacinia
Order Lepidoptera   Butterflies
Family Nymphalidae   Brush-footed Butterflies

The bordered patch is a medium-sized butterfly that is closely related to the checkerspot butterflies. It has a wingspan of about 1.5–2.0 inches. On top, the wings are dark with a band of orange and small white spots on the edges. The caterpillars feed primarily on sunflower, cowpen daisy, and giant ragweed.

Queen
Danaus gilippus
Order Lepidoptera   Butterflies
Family Nymphalidae   Brush-footed Butterflies

The queen is more brownish-red in color than its close relative the monarch. Also, the upper surface of the wings of the queen lacks the black veins found on the monarch. Like monarchs, the caterpillars feed on milkweed plants. Their toxicity to predators varies depending on the species of milkweed they eat.

Monarch
Danaus plexippus
Order Lepidoptera   Butterflies
Family Nymphalidae   Brush-footed Butterflies

The monarch is orange with black wing veins and white spots along the edges of the wings. Monarchs migrate in large numbers through the Hill Country in the spring and fall on the way from and to their winter home in the mountains of central Mexico. Monarch caterpillars, which have white, yellow, and black stripes, feed on milkweed plants, sequestering toxic chemicals from the plant that in turn makes the butterfly toxic to predators. The bright orange coloration of adults enables predators to easily recognize and avoid this toxic prey item.

Snout Butterfly
Libytheana carinenta
Order Lepidoptera   Butterflies
Family Nymphalidae   Brush-footed Butterflies

The snout butterfly is easily recognized by the mouthparts that extend forward with the antennae from the head. The wings are brown and orange with white spots. The tip of the forewing is squared off. The underside is brown, and the butterfly looks like a dead leaf when it lands. This species undergoes mass migrations when the population explodes, usually in the summer or fall. Driving during these periods results in dozens of smashed butterflies on the windshield and grill of your car. The caterpillars are green and feed on hackberry leaves.

Pipevine Swallowtail
Battus philenor
Order Lepidoptera   Butterflies
Family Papilionidae   Swallowtail Butterflies

The pipevine swallowtail has iridescent blue-black wings. The underside of the hind wings has seven orange spots along the edge. The hind wings have small extensions that resemble the tail of a swallow. The caterpillars are either black or reddish-orange, depending on the temperature at which they develop (warmer temperatures produce the reddish-orange caterpillars). The caterpillars feed on pipevine plants but often crawl up taller vegetation, possibly to keep cool. Several other species of swallowtails also occur in the Hill Country.

Praying Mantid
Order Mantodea   Mantids

The praying mantid (or mantis) is a well-known predatory insect, and several species occur in Texas. The front legs are modified for grasping prey. The body is green to brown and fairly slender, with the wings folded up along the back in adults. The eggs are laid on a stem in a frothy material, which then hardens and looks somewhat like a petrified caterpillar stuck to the stem.

Scorpionflies
Order Mecoptera   Scorpionflies and Hanging Flies
Family Panorpidae   Common Scorpionflies

Scorpionflies are unusual insects with an abdomen that curls up like a scorpion's. They do not sting or bite humans. They have four wings that are striped with black and a long beak below the eyes. The species most commonly seen in the Hill Country is *Panorpa nuptialis*. They are usually seen in the fall and feed primarily on dead insects.

Green Lacewing
Order Neuroptera   Nerve-winged Insects
Family Chrysopidae   Green Lacewings

The green lacewing is about ½ inch long, with a green body and transparent wings with green veins, held back over the body. Due to the green color, they are most easily seen when perched on window screens or buildings. They are predators on other insects and are considered beneficial to humans because the larvae can consume large numbers of aphids and other plant pests.

Dobsonflies
Order Neuroptera   Nerve-winged Insects
Family Corydalidae   Dobsonflies

These huge insects are distinctive in both the adult and larval stages. The larvae, also known as hellgrammites, live in fast-flowing rocky areas of streams, where they prey on insects and other small animals. The adults look somewhat like dragonflies, but with wings folded back and with large pinching mouthparts. They are most commonly seen attracted to lights near rivers.

Antlions
Myrmeleon spp.
Order Neuroptera   Nerve-winged Insects
Family Myrmeleontidae   Antlions

Antlion larvae build funnel-shaped pits in sandy soils. The larvae, also called doodlebugs, hide in the sand at the base of the pit. When ants wander into the pit, the steep sides with loose sand prevent them from crawling out, and the antlion may throw sand onto the sides to further prevent their escape. The antlion larvae have large mandibles used to grab the prey. Adult antlions have clear wings and look like gray damselflies but have obvious antennae with a small club on the end.

Dragonfly

Dragonflies and Damselflies
Order Odonata

Dragonflies and damselflies are common insects near water but are occasionally also seen far from water. Dragonflies are larger bodied, with rapid flight and wings held open when perched. Damselflies are slender, with more bouncy flight and wings held together when perched. Both have aquatic larval stages. Larvae and adults are predators, mostly on other insects.

Grasshopper

Grasshoppers, Crickets, and Katydids
Order Orthoptera

Grasshoppers and their relatives are among the most abundant large insects in the Hill Country and are among the noisiest. All have fairly long hind legs for jumping and wings folded over the back. They create sounds by rubbing their wings together. Tree crickets create a high-pitched whine or trill in the summer evenings. Follow this sound to its source, and you will find a small green cricket with long antennae. Field crickets create a chirping sound at night. They are brown to black and usually seen on the ground or in buildings. Loud daytime sounds are more likely those of the cicadas (Order Hemiptera).

Giant Walkingstick

Walkingsticks
Order Phasmida (Phasmatodea)

Walkingsticks are easily recognized by their long, thin body and legs. They feed on the leaves of trees and shrubs. There are 16 species in Texas, one of which is the longest insect in the United States, the giant walkingstick (*Megaphasma dentricus*), that can reach 7 inches long.

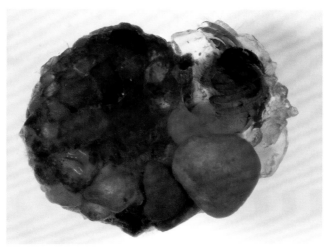

Snail-case caddisfly

Caddisflies
Order Trichoptera

The larvae of caddisflies occur in many streams of the Hill Country. They build a case made of sand, small rocks, leaf fragments, or sticks. The snail-case caddisfly (*Helicopsyche* spp.) uses tiny stones to build a spiraling case with the shape of a snail. Some caddisfly larvae create silk nets to capture food from the stream current. The larva itself looks somewhat like a pale caterpillar if removed from the case, and caddisflies are related to butterflies and moths. After pupating in the water, the adult caddisfly emerges to the terrestrial world for a short life as a reproductive adult. The adults resemble small moths with hairy wings.

CHAPTER ELEVEN **State Parks and Other Natural Areas**

S OME OF THE SIGNIFICANT NATURAL AREAS in the Hill Country, where one can observe many of the species described in this book, are listed here. Texas Parks and Wildlife Department (TPWD) has also developed "wildlife trails" that list sites throughout the state that are open for wildlife observation. In the Hill Country, these are known as the Heart of Texas Wildlife Trails (East and West), and information on them is available on the TPWD website.

Barton Springs

Austin City Parks

The city of Austin has several parks that provide access to the nature of the Hill Country. Barton Springs and Barton Creek Wilderness Park (1,021 acres) are located in the center of Austin. The spring-fed swimming pool at Zilker Park is also the habitat for the endangered Barton Springs salamander. Upstream, the Barton Creek Greenbelt extends southwest along the creek and includes the Barton Creek Wilderness Park. A 7-mile trail for hiking or mountain biking follows the creek. The Bull Creek Greenbelt is located on the north side of Austin and has 3.5 miles of trail. Emma Long Metropolitan Park is on the west side of town along the Lake Austin section of the Colorado River.

Balcones Canyonlands National Wildlife Refuge (20,000 acres)

Balcones Canyonlands is the largest public natural area in the Hill Country, but only a small portion is open to public use. The primary purpose of the refuge is to protect the nesting habitat of the Golden-cheeked Warbler and Black-capped Vireo, as well as the water quality of the area. The proposed refuge size is 46,000 acres, but only about half has been purchased as of 2013. Hiking trails are open to the public at the Doeskin Ranch tract (5 miles of trails), Warbler Vista (2.5 miles), and at the headquarters (0.4 mile).

Blanco State Park (105 acres)

This small park on the Blanco River, within the town of Blanco, is popular for fishing and swimming. Two small dams create shallow reservoirs. This park has extensive access to the riverbanks and a short nature trail along the river. There is a small campground and a wildlife viewing station with bird feeders.

Canyon Lake Dinosaur Tracks

A large set of dinosaur tracks is easily observed at this location on the south side of Canyon Lake. The tracks are part of the Heritage Museum of the Texas Hill Country. There is an entrance fee to see the tracks.

Canyon Lake Gorge

In 2002, 34 inches of rain in the upper Guadalupe River basin filled the Canyon Lake reservoir to overflowing, carving a mile-long canyon through the limestone below the spillway. A spring-fed stream flows through the gorge. The gorge is open only to guided tours, which are available on Saturday mornings by reservation.

(Photograph by LCRA)

Canyon of the Eagles Nature Park (940 acres)

This park, owned by the Lower Colorado River Authority (LCRA), is also home to a resort called Canyon of the Eagles. The area is named for the wintering Bald Eagles along the lakeshore. Boat tours are available from Vanishing Texas River Cruises to observe the eagles and other wildlife. The park has 14 miles of hiking trails.

Cibolo Nature Center (100 acres)

This city park is located off State Highway 46 on the east side of Boerne. The tall bald cypress trees along Cibolo Creek are a highlight of the park. There are several hiking trails, a visitor center, and a replica of a dinosaur trackway. There are also trails through a restored tallgrass prairie and a boardwalk over a restored marsh.

Gorman Falls at Colorado Bend State Park

Colorado Bend State Park (5,328 acres)

Colorado Bend is located in the northern part of the Hill Country, 28 miles west of Lampasas. This park is one of the more remote state parks in the Hill Country, with a 6-mile-long gravel road into the park. The Colorado River flows along the north and east sides of the park. Gorman Falls is a 60-foot waterfall located near the river, at the end of a 1-mile hike. The spring water that flows over the waterfall is rich in calcium carbonate and has created travertine formations below the falls (travertine is limestone deposited by evaporation). The area around the waterfall is densely vegetated, and the cliffs behind the falls are covered with moss and ferns. There is no swimming in the stream or waterfall, but swimming is available in the Colorado River. The park has 30 miles of hiking trails, most of which are also open to mountain bikers.

Enchanted Rock State Natural Area (1,644 acres)

Enchanted Rock is a large granite dome that rises 425 feet above the surrounding ground and covers 640 acres. It is located north of Fredericksburg. Several smaller outcrops surround it. The top of Enchanted Rock contains small depressions, called vernal pools, which fill with water after rains. These pools are home to some unique plants and to fairy shrimp, small crustaceans that swim upside down. The fairy shrimp survive drought periods in a resistant egg stage and occur only in aquatic habitats that lack fish. Enchanted Rock is an exfoliation dome, meaning that layers of rock break off the surface of the dome. These layers can be seen in various states of cracking and sliding down the dome. Because new rock surface is continually uncovered, Enchanted Rock is an excellent place to observe succession, the change in an ecological community following a disturbance. Lichens are the first pioneer species to colonize the rock, followed by small plants, grasses and shrubs, and eventually trees. Hiking trails lead to the top of the rock as well as around it. Camping includes walk-in sites and backpacking.

Garner State Park (1,774 acres)

Garner is located in the southwestern Hill Country, 30 miles north of Uvalde, on the crystal-clear Frio River. It is the most popular park in Texas for camping. The park contains rugged hillsides dominated by Ashe juniper and flat areas along the river with bald cypress, pecan, and sycamore. Axis deer are common in the park. Mexican fern (*Anemia mexicana*) is abundant on the rocky hillsides. Several miles of hiking trails lead over the hills and through the canyons of the southern half of the park, and a mountain bike trail traverses the open woodlands of the floodplain in the northern half of the park. There is a bird-viewing blind near the visitor center.

Government Canyon State Natural Area (8,624 acres)

Government Canyon is located on the west side of San Antonio. Ashe juniper, Texas mountain laurel, agarita, and live oak are common along the trails of the backcountry region, with mesquite common in the more open "frontcountry" region. The creeks here flow on an intermittent basis (after rains). A portion of the natural area is closed to use from March through August to protect breeding habitat for Golden-cheeked Warblers, but there are plenty of other trails open during these months. The protected area also contains caves that are home to eight species of endangered invertebrates. This large natural area, most of which is part of the Edwards Aquifer recharge zone, helps protect the drinking water supply for San Antonio. The natural area opened in 2005, and a campground with walk-in tent sites is now open. Most trails are open to mountain biking as well as hiking.

Guadalupe River State Park

Guadalupe River State Park (1,939 acres) and Honey Creek State Natural Area (2,294 acres)

These adjoining parks are about 30 miles north of downtown San Antonio. Access to Honey Creek is restricted, but guided hikes are available on Saturday mornings. This is a unique opportunity to see a beautiful spring-fed stream in pristine condition. Guadalupe River State Park features a long, high cliff along the river. The riverbank is lined by large bald cypress trees, as well as sycamore and pecan. The upland areas contain grassland openings and wooded areas of Ashe juniper, live oak, Texas persimmon, and agarita. The river is popular for swimming, tubing, and kayaking (dependent on flow conditions). The park has several miles of hiking/mountain biking/equestrian trails and three campgrounds. A bird blind and nature center are located near the river.

Hill Country State Natural Area (5,370 acres)

Hill Country SNA is 50 miles northwest of San Antonio. Like many state parks, it was once a ranch and maintains facilities for those with horses. The landscape is open savannas with some dense juniper woodlands on the hillsides. A small creek flows through the park. Camping is primitive with no potable water. There are 40 miles of trails for hiking, horseback riding, and mountain biking.

Inks Lake State Park (1,201 acres)

Inks Lake is located 9 miles west of Burnet in the Llano Uplift region. Exposures of pink Valley Spring Gneiss are abundant around the reservoir and throughout the park. The Devil's Waterhole is a scenic corner of the lake with rock walls. Small vernal pools are present on the gneiss outcrops, similar to the pools at Enchanted Rock. The park has 7.5 miles of hiking trails and a very large campground with cabins.

The Blue Hole

Jacob's Well Natural Area (80 acres)
and Blue Hole Regional Park (126 acres)

Jacob's Well, west of Wimberley, is an artesian spring and the entrance to an underwater cave. The well extends straight down 30 feet, but this is only the beginning of an extensive underwater cave system. Public tours are available on Saturday mornings. Swimming is permitted after the tour, but recreational diving is no longer allowed due to the deaths of several divers in the cave. The Blue Hole is located on the east side of Wimberley, downstream from Jacob's Well. A large swimming area shaded by huge bald cypress trees is the central attraction. The park is open daily, but swimming is allowed only during the warmer months.

Kerr Wildlife Management Area (6,493 acres)

Kerr WMA is located in the central Hill Country, 13 miles west of
Hunt on FM 1340. Its primary use is for research into game management methods. The vegetation on the WMA is a mixture of grassy
pastures and woods of live oak and Ashe juniper. The area is often
visited by birders in the spring because of breeding populations of
Black-capped Vireos and Golden-cheeked Warblers. Information on
where to view these birds is available at the headquarters. There is a
4-mile paved driving tour, with several stops with signs that explain
principles of game management in the Hill Country. The Spring Trap
Trail (1 mile) is the only hiking trail on the WMA. Biking is allowed on
the paved road. The WMA does not include any land along the nearby
north fork of the Guadalupe River.

(Photograph by Earl Nottingham, TPWD)

Kickapoo Cavern State Park (6,368 acres)

Kickapoo Cavern State Park is a fairly remote park in the western Hill Country north of Brackettville. The park is open only Friday to Monday. Saturday cave tours are by reservation only. Since the cave is undeveloped, the cave tours require good hiking abilities. Bats use one of the caves and can also be observed when they are present. The park includes 14 miles of mountain biking trails, 18 miles of hiking trails, and a primitive campground.

Kreutzberg Canyon Natural Area (117 acres)

This park, which opened in 2012, is located on Mark Twain Drive about 10 miles northeast of Boerne. The landscape ranges from upland oak and juniper woodlands to the pecans and bald cypress along the Guadalupe River. Three miles of trails are available for hiking or mountain biking. There are three bird blinds for observation and photography.

Landa Park (51 acres)

Landa Park is a large city park in the heart of New Braunfels. In addition to the playgrounds, swimming pool, and golf course, there are two natural features of interest. The first is Comal Springs, the largest springs in Texas. There are two large springs near the north end of the park, as well as many that lie under the lake. These provide the flow for the Comal River. The second natural attraction is Panther Canyon Nature Trail, which begins near the large spring outlet and follows the bottom of the canyon for 0.8 mile until it ends in a residential neighborhood.

Longhorn Cavern State Park (646 acres)

Longhorn Cavern is located near Marble Falls, just south of Inks Lake State Park. The cave is formed in limestone that is 450 to 500 million years old. Tours of the cavern are available daily. The park is day use only and has just over a mile of nature trails.

Lost Maples State Natural Area (2,174 acres)

Lost Maples, located near Vanderpool in the south-central part of the Hill Country, is one of the most popular parks in Texas. The main attraction is the bigtooth maples, which turn red in November, drawing hundreds of visitors on the weekends. The park posts a weekly update on the fall foliage on its website. Steep-walled canyons and beautiful spring-fed streams make this one of the best parks for hiking in the Hill Country during any season. There are more than 10 miles of hiking trails, several backpacking sites, and a small campground.

McKinney Falls State Park (726 acres)

McKinney Falls is on the southeast side of Austin. There are two main waterfalls along Onion Creek, which is lined by bald cypress, sycamore, and pecan trees. A huge bald cypress tree, estimated at more than 500 years old, is located below a bridge on the Rock Shelter Interpretive Trail. The rock shelter on this trail was used by American Indians for centuries. Two other trails are open for hiking and mountain biking; each is about 3 miles long. Swimming is allowed below both falls.

Natural Bridge Caverns

Natural Bridge Caverns is located between New Braunfels and San Antonio. The caverns have been designated as a National Natural Landmark because of many unique formations. These are the largest commercial caverns in Texas. The name of the caverns comes from a natural stone bridge near the entrance. A variety of guided tours are available.

Pedernales Falls State Park (5,212 acres)

Pedernales Falls is in the central Hill Country between Johnson City and Austin. The park contains several miles of the Pedernales River, including a series of low waterfalls and rapids. Large areas of tilted limestone are kept clear of vegetation by flash floods. Texas earless lizards are common on the rocks. Swimming is allowed in the river but not at the falls. An excellent bird blind attracts a variety of birds for close viewing and photography. There are developed and primitive campgrounds. The Twin Falls Trail leads to an overlook of a fern-covered canyon. Other trails lead through juniper-dominated uplands and across spring-fed creeks lined with bald cypress and sycamore. Trails are available for hiking, mountain biking, and horseback riding. The name of the park is usually pronounced "purd-in-ALICE."

(Photograph by Susan Sander)

Riverside Nature Center, Kerrville

This urban wildlife sanctuary is located near downtown Kerrville. It is a nonprofit organization with a mission to promote understanding and stewardship of the natural resources of the Hill Country. A nature trail leads through an arboretum of more than 70 species of trees and shrubs that are native to the Hill Country, as well as areas with wildflowers, cacti, and a butterfly garden. A visitor center and educational programs are also available. Access is free. It provides a trailhead for Kerrville's River Trail.

Eisenhower Park

San Antonio City Parks

The northern part of San Antonio is within the Hill Country, and there are several natural areas worth visiting. Friedrich Wilderness Park (633 acres) and Eisenhower Park (317 acres) are both excellent destinations for hiking. McAllister Park (984 acres) is popular for mountain biking. Smaller parks with short hiking trails include Crownridge Canyon, Stone Oak, and Comanche Lookout.

Purgatory Creek Natural Area

San Marcos City Parks

The city of San Marcos has several natural areas open to hiking and mountain biking. Trail maps are available on the website of the San Marcos Greenbelt Alliance. Purgatory Creek Natural Area (570 acres) is located on the west side of San Marcos. The park contains uplands with mesquite, rocky hillsides, and an intermittent creek. Spring Lake Preserve (251 acres) is a hidden gem in the city of San Marcos. It preserves a portion of the watershed that flows into Spring Lake, the headwaters of the San Marcos River and home to San Marcos Springs. The preserve is dominated by Ashe juniper and live oak–covered hills. Texas mountain laurel, mesquite, and prickly pear are also very common. About 1 mile from the entrance, there is a small cattail-bordered pond with a dock. A portion of the park is closed in the spring to protect the nesting areas of Golden-cheeked Warblers. Parking is located at the Texas Rivers Center at Texas State University, near the springs. The trail begins across the street near the golf course (follow the blue rectangles).

South Llano River State Park (2,600 acres)

Located in Junction, this park is more arid than many of the other parks in the Hill Country. Average annual precipitation is only about 23 inches. The South Llano River is one of the most pristine rivers in Texas and has become popular for recreation. The woodlands along the river are home to a large roosting population of Wild Turkeys. The park has four wildlife observation blinds that attract a wide variety of birds for close viewing. There is a campground and a network of 25 miles of hiking and biking trails.

Hamilton Pool

Travis County Parks

Travis County contains several parks, including county parks and several owned by the Lower Colorado River Authority (LCRA). Reimers Ranch (2,427 acres) is popular for swimming and mountain biking and protects land along the Pedernales River. Pace Bend (1,368 acres) is popular for its shoreline on Lake Travis. Turkey Bend (400 acres) is also on Lake Travis. Hamilton Pool Nature Preserve (232 acres) protects a natural pool that lies under a 50-foot waterfall. This park is very popular in the summer for swimming and has limited parking. Next to Hamilton Pool is Westcave Preserve, which protects a waterfall and caves and is open by guided tour only.

References

GENERAL REFERENCES AND INTRODUCTION

Allred, L. 2009. *Enchanted Rock: A Natural and Human History.* Austin: University of Texas Press.

Hawthorne, J. M. 1990. *Dinosaur Track-Bearing Strata of the Lampasas Cut Plain and Edwards Plateau, Texas.* Baylor Geological Studies, Bulletin No. 49. Waco: Baylor University.

La Vere, David. 2004. *The Texas Indians.* College Station: Texas A&M University Press.

Smeins, F., S. Fuhlendorf, and C. Taylor Jr. 1997. *Environmental and Land Use Changes: A Long-Term Perspective.* College Station: Juniper Ecology and Management Symposium, Texas A&M University.

Spearing, D. 1991. *Roadside Geology of Texas.* Missoula, MT: Mountain Press Publishing.

Stanley, J. 2009. *Hill Country Landowner's Guide.* College Station: Texas A&M University Press.

Swanson, E. R. 1995. *Geo-Texas.* College Station: Texas A&M University Press.

Toomey, R. S. 1994. "Vertebrate Paleontology of Texas Caves." In *The Caves and Karst of Texas,* ed. W. R. Elliott and G. Veni, 51–68. Huntsville, AL: National Speleological Society.

Veni, G. 1994. "Hydrogeology and Evolution of Caves and Karst in the Southwestern Edwards Plateau, Texas." In *The Caves and Karst of Texas,* ed. W. R. Elliott and G. Veni, 13–30. Huntsville, AL: National Speleological Society.

PLANTS

Cox, P. W., and P. Leslie. 1997. *Texas Trees, a Friendly Guide.* San Antonio: Corona Publishing.

Diggs, G. M., Jr., B. L. Lipscomb, and R. J. O'Kennon. 1999. *Shinner & Mahler's Illustrated Flora of North Central Texas.* Fort Worth: Botanical Research Institute of Texas.

Enquist, M. 1987. *Wildflowers of the Texas Hill Country.* Austin: Lone Star Botanical.

Flora of North America Editorial Committee, eds. 1993+. *Flora of North America North of Mexico.* Available at www.efloras.org.

Loflin, B., and S. Loflin. 2006. *Grasses of the Texas Hill Country.* College Station: Texas A&M University Press.

Loughmiller, C., and L. Loughmiller. 2006. *Texas Wildflowers, a Field Guide.* Rev. ed. Austin: University of Texas Press.

Lynch, D. 1981. *Native and Naturalized Woody Plants of Austin and the Hill Country.* Austin: St. Edward's University.

Powell, A. M., J. F. Weedin, and S. A. Powell. 2008. *Cacti of Texas.* Lubbock: Texas Tech University Press.

Wrede, J. 2010. *Trees, Shrubs, and Vines of the Texas Hill Country.* 2nd ed. College Station: Texas A&M University Press.

Yarborough, S. C., and A. M. Powell. 2002. *Ferns and Fern Allies of the Trans-Pecos and Adjacent Areas.* Lubbock: Texas Tech University Press.

FUNGI AND LICHENS

Brodo, I. M., S. D. Sharnoff, and S. Sharnoff. 2001. *Lichens of North America.* New Haven, CT: Yale University Press.

Metzler, S., and V. Metzler. 1992. *Texas Mushrooms: A Field Guide.* Austin: University of Texas Press.

BIRDS

Benson, K. L. P., and K. A. Arnold. 2001. *The Texas Breeding Bird Atlas.* College Station and Corpus Christi: Texas A&M University System. Available at http://txtbba.tamu.edu.

Kaufman, K. 2000. *Birds of North America.* New York: Houghton Mifflin.

Kutac, E. A., and S. C. Caran. 1994. *Birds and Other Wildlife of South Central Texas.* Austin: University of Texas Press.

Lockwood, M. W. 2001. *Birds of the Texas Hill Country.* Austin: University of Texas Press.

Rylander, K. 2002. *The Behavior of Texas Birds.* Austin: University of Texas Press.

MAMMALS

Goetze, J. R. 1998. *The Mammals of the Edwards Plateau, Texas.* Museum of Texas Tech University, Special Publications No. 41. Lubbock: Texas Tech University.

Kays, R. W., and D. E. Wilson. 2002. *Mammals of North America.* Princeton, NJ: Princeton University Press.

Kutac, E. A., and S. C. Caran. 1994. *Birds and Other Wildlife of South Central Texas.* Austin: University of Texas Press.

McGee, B. K., and R. W. Manning. 2000. *Mammals of Lost Maples State Natural Area, Texas.* Occasional Papers of the Museum of Texas Tech University, no. 198. Lubbock: Texas Tech University.

Schmidly, D. J. 2004. *The Mammals of Texas.* Rev. ed. Austin: University of Texas Press.

Wolverton, S., J. H. Kennedy, and J. D. Cornelius. 2007. "A Paleozoological Perspective on White-tailed Deer (*Odocoileus virginianus texana*) Population Density and Body Size in Central Texas." *Environmental Management* 39:545–52.

REPTILES AND AMPHIBIANS

Conant, R., and J. T. Collins. 1998. *A Field Guide to Reptiles and Amphibians: Eastern and Central North America.* 3rd ed. New York: Houghton Mifflin.

Dixon, J. R. 2000. *Amphibians and Reptiles of Texas.* 2nd ed. College Station: Texas A&M University Press.

Elliot, L., C. Gerhardt, and C. Davidson. 2009. *The Frogs and Toads of North America.* New York: Houghton Mifflin Harcourt.

Kutac, E. A., and S. C. Caran. 1994. *Birds and Other Wildlife of South Central Texas.* Austin: University of Texas Press.

Tipton, B. L., T. L. Hibbits, T. D. Hibbits, T. J. Hibbits, and T. Laduc. 2012. *Texas Amphibians: A Field Guide.* Austin: University of Texas Press.

Werler, J. E., and J. R. Dixon. 2000. *Texas Snakes.* Austin: University of Texas Press.

FISH

Page, L. M., and B. M. Burr. 1991. *A Field Guide to Freshwater Fishes North of Mexico.* New York: Houghton Mifflin.

Thomas, C., T. H. Bonner, and B. G. Whiteside. 2007. *Freshwater Fishes of Texas.* College Station: Texas A&M University Press.

INVERTEBRATES

Drees, B. M., and J. A. Jackman. 1998. *A Field Guide to Common Texas Insects.* Houston: Gulf Publishing.

Jackman, J. A. 1999. *A Field Guide to Spiders and Scorpions of Texas.* Houston: Gulf Publishing.

Kattes, D. H. 2009. *Insects of Texas: A Practical Guide.* College Station: Texas A&M University Press.

Thorp, J. H., and D. C. Rogers. 2011. *Field Guide to Freshwater Invertebrates of North America.* Boston: Academic Press.

Index

Acer grandidentatum, 24
Acer negundo, 25
Acris (crepitans) blanchardi, 245
addax antelope, 195
Adiantum capillus-veneris, 117
Agalinis spp., 88
agarita, 23, 33, 315–16
Agelaius phoeniceus, 188
Agkistrodon contortrix, 238
Agkistrodon piscivorous, 238
Agraulis vanillae, 289
algae, 125–26, 129, 132, 221, 266, 278
Allowissadula holosericea, 62
Aloysia gratissima, 57
Ameiurus spp., 258
amphipod, Peck's cave, 14
Anaxyrus spp., 244
Andropogon glomeratus, 112
Anemia mexicana, 116, 314
anemone: desert, 99; two-flow-
    ered, 99
Anemone spp., 99
anole, green, 231
Anolis carolinensis, 231
ant: harvester, 227, 286; red
    imported fire, 227, 287
antelope: addax, 195; blackbuck, 195
antelope-horns, 65
antlion, 299
Antrostomus carolinensis, 157
Apache Indians, 15
Apalone spinifera, 222
Aphelocoma californica, 169
Aphonopelma spp., 268
Apis mellifera, 284

aquifer, 8, 10–14, 16–20, 242;
    Edwards, 10–13, 19, 242, 315; lev-
    els, 10, 18–19; protection, 12,
    19–20
Arbutus xalapensis, 38
Archilochus spp., 159
Ardea herodias, 142
Argemone spp., 97
Argiope aurantia, 267
armadillo, nine-banded, 205
Arundo donax, 113
Asclepias asperula, 65
ash, Texas, 23, 51
Aspidoscelis spp., 233
Asterocampa celtis, 290
Austin City Parks, 305
Axis axis, 213

Baccharis neglecta, 32
Baeolophus atricristatus, 173
Baiomys taylori, 202
Balcones Canyonlands, 1, 306
Balcones Canyonlands National
    Wildlife Refuge, 20, 139, 306
Balcones Escarpment, 18
Balcones Fault Zone, 6, 8
Bamberger Ranch Preserve, 17
barberry, Texas, 33
Barton Springs, 8, 11, 305
bass: Guadalupe, 13, 262; large-
    mouth, 262; smallmouth, 262;
    striped, 260, 262; white, 260, 262
Bassariscus astutus, 208
bat, 195, 197, 321; Mexican free-
    tailed, 197

brake, 119; ovate maiden, 120; southern maidenhair, 117; wooly lip, 118
ferret, black-footed, 17
Finch, House, 192
fire ants, red imported, 287
firewheel, 60
flash floods, 7, 10, 328
flower structures (parts), 59
fly: black, 278; leafminer, 277
Flycatcher: Scissor-tailed, 166; Vermilion, 164
Forestiera pubescens, 50
fossils, 4–7
fox: common gray, 212; red, 212
Fraxinus texensis, 51
Friedrich Wilderness Park, 330
Friesenhahn Cave, 7
fritillary, Gulf, 289
frog: Blanchard's cricket, 245; cliff chirping, 241, 249; gray/Cope's gray treefrog, 246; green treefrog, 247; leopard, 251; Rio Grande leopard, 251; spotted chorus, 248; Strecker's chorus, 248
frostweed, 93, 277
fungi, 44, 129–32, 336; shelf, 130

Gaillardia pulchella, 60
galls, 45, 285
gambusia: largespring, 13; San Marcos, 13
Gambusia affinis, 259
Gambusia geiseri, 13
Gambusia georgei, 13
gar: longnose, 254; spotted, 254
Garner State Park, 21, 116, 314
Garrya ovata, 48
gartersnake: black-necked, 236; checkered, 236

Gastrophryne olivacea, 250
gecko, Mediterranean, 225
Geococcyx californianus, 154
geology, 1, 4–9
Geomys texensis, 198
German immigrants, 16
Gerrhonotus infernalis, 223
Glandularia bipinnatifida, 90
Gnathamiterme spp., 288
Goldfinch: American, 193; Lesser, 193
gopher, pocket, 198
Gorman Cave, 9
gourd: buffalo, 71; stinking, 71
Government Canyon State Natural Area, 315
Grackle: Common, 190; Great-tailed, 190
granite, 4, 106, 119, 124
grape: mountain, 111; mustang, 111; sand, 111; summer, 111
Graptemys caglei, 219
Graptemys versa, 219
grasses, 101, 112–15, 204, 206, 313, 336
grasshopper, 301
greenbrier, 109
greensnake, rough, 236
groundwater conservation districts, 12
Guadalupe River, 11, 16, 253, 320, 322
Guadalupe River State Park, 316
Gull: Franklin's, 149; Herring, 149; Laughing, 149; Ring-billed, 149

hackberry, 34, 290, 294; netleaf, 22, 34; sugar, 34
Haemorhous mexicanus, 192
Hamilton Pool Nature Preserve, 333
harvestmen, 269
Hawk: Red-shouldered, 147; Red-tailed, 147

bush, 23, 26

sunfish: bluegill, 261; green, 261; longear, 261; redbreast, 261; red-ear, 261

sunflower, 59, 67, 69, 291; common, 67; Maximilian's, 67

Sus scrofa, 216

Swallow: Barn, 171; Cave, 171; Cliff, 171

swallowtail butterfly, pipevine, 295

Swift, Chimney, 158

sycamore, American, 23, 52, 314, 316, 326, 328

Sylvilagus spp., 207

Tadarida brasiliensis, 197

tallow, Chinese, 22, 39

tarantula, 268

tasajillo, 103

Taxodium distichum, 36

Teloschistes spp., 136

termite, 250, 288; desert, 288; subterranean, 288

Thamnophis spp., 236

Thelypteris ovata, 120

thistle, Texas, 79

Thryomanes bewickii, 176

Thryothorus ludovicianus, 175

Tillandsia recurvata, 121

Tillandsia usneoides, 122

Titmouse, Black-crested, 173

toad: coastal plain, 244; Couch's spadefoot, 252; green, 244; Gulf Coast, 244; horned, 227; red-spotted, 244; Texas, 244; western narrow-mouthed, 250; Woodhouse's, 244

Tonkawa Indians, 15

Towhee: Canyon, 184; Eastern, 184; Green-tailed, 184; Spotted, 184

Toxicodendron radicans, 108

Trachemys scripta, 220

Tradescantia, 83

Travis County Parks, 333

treefrog: gray, 246; green, 247

Trogloglanis pattersoni, 13

trout, rainbow, 253

Turdus migratorius, 179

Turkey, Wild, 49, 140, 332

Turkey Bend, 333

Turk's cap, 63

turtle: Cagle's map, 219; eastern musk, 221; Mississippi mud, 221; red-eared slider, 220; snapping, 218; spiny softshell, 222; Texas map, 219; Texas river cooter, 220; yellow mud, 221

Tyrannus forficatus, 166

Tyrannus verticalis, 165

Ulmus spp., 56

Ungnadia speciosa, 55

Urocyon cinereoargenteus, 212

Urosaurus ornatus, 230

Usnea spp., 137

venomous snakes, 217, 234, 237–40

Verbascum thapsus, 74

verbena, prairie, 90

Verbesina encelioides, 68

Verbesina virginica, 93

vines, 59, 83, 100–101, 108–11, 336

vipers, 237–40

Vireo, Black-capped, 47, 139, 167, 306, 320

Vireo atricapilla, 167

Virginia striatula, 237

Vitis spp., 111

Vulpes vulpes, 212

Vulture: Black, 145; Turkey, 145–46